Nobel and Lasker Laureates of Chinese Descent
In Literature and Science

炎黃子孫的諾貝爾和拉斯克獎得主:於文學與科學

Edited by

Todd S. Ing, Keith K. Lau, Joseph M. Chan,
Hon-Lok Tang, Angela T. Hadsell
and Laurence K. Chan

With the assistance of Cheryl D. Lau

Royalties derived from this book are donated to the
University of Hong Kong and the Chinese University of Hong Kong.

World Scientific

NEW JERSEY · LONDON · SINGAPORE · BEIJING · SHANGHAI · HONG KONG · TAIPEI · CHENNAI · TOKYO

Published by

World Scientific Publishing Co. Pte. Ltd.

5 Toh Tuck Link, Singapore 596224

USA office: 27 Warren Street, Suite 401-402, Hackensack, NJ 07601

UK office: 57 Shelton Street, Covent Garden, London WC2H 9HE

Library of Congress Cataloging-in-Publication Data
Names: Ing, Todd S., editor. | Lau, Cheryl D.
Title: Nobel and Lasker laureates of chinese descent : in literature and science /
  edited by Todd S. Ing [and five others] ; with the assistance of Cheryl D. Lau.
Description: New Jersey : World Scientific, 2019. | Includes bibliographical references and index.
Identifiers: LCCN 2018013166| ISBN 9789814704601 (hardcover : alk. paper) |
  ISBN 9814704601 (hardcover : alk. paper) | ISBN 9789814704618 (pbk. : alk. paper) |
  ISBN 981470461X (pbk. : alk. paper)
Subjects: LCSH: Physicists--China. | Chemists--China. | Authors--China. | Medical scientists--China. |
  Nobel Prize winners--China. | Albert Lasker Awards.
Classification: LCC QC9.C6 N63 2018 | DDC 001.4/409239951--dc23
LC record available at https://lccn.loc.gov/2018013166

British Library Cataloguing-in-Publication Data
A catalogue record for this book is available from the British Library.

Copyright © 2019 by World Scientific Publishing Co. Pte. Ltd.

All rights reserved. This book, or parts thereof, may not be reproduced in any form or by any means, electronic or mechanical, including photocopying, recording or any information storage and retrieval system now known or to be invented, without written permission from the publisher.

For photocopying of material in this volume, please pay a copying fee through the Copyright Clearance Center, Inc., 222 Rosewood Drive, Danvers, MA 01923, USA. In this case permission to photocopy is not required from the publisher.

For any available supplementary material, please visit
http://www.worldscientific.com/worldscibooks/10.1142/9716#t=suppl

Preface

We wish to dedicate this book to two exemplary masters.

A. Brother Brendan Dunne

桃李滿天下 (Peaches and Plums All over the World).

N.B. In classic Chinese, students and trainees are likened to peaches and plums.

This book honors Brother Brendan Dunne, FSC (畢列登修士) 1914–1998, a De La Salle Brother of Irish descent, of St. Joseph's College, a prestigious, English-centric, secondary boys' school run by the De La Salle Brothers in Hong Kong. Brother Brendan devoted his life to educate students from Hong Kong, with Laureate Charles Kao being one of his stellar pupils. The above Chinese accolade fits Brother Brendan's outstanding achievements to a T. (Photo courtesy of Dr. Peter H. T. Wu.)

B. **Professor Lap-Chee Tsui** (徐立之). BSc (CUHK), MPhil (CUHK), PhD (University of Pittsburgh), OC, GBM, FRCP, FRS, FRSC. The 14th Vice-Chancellor and President of the University of Hong Kong, 2002–2014.

This book also honors Professor Tsui, whose groundbreaking work in human genetics led to the discovery of the Cystic Fibrosis Transmembrane Conductance Regulator (CFTR) gene, which leads to cystic fibrosis when mutated. In addition, he has made significant contributions to the study of the human genome, particularly the characterization of chromosome 7 and the identification of other disease genes.

Reference

Tsui LC, Buchwald M, Barker D, *et al.* (1985) Cystic fibrosis locus defined by a genetically linked polymorphic DNA marker. *Science* **29:230** (4729): 1054–1057.

To Leave Footprints Behind on the Sands of Time!

By trumpeting the momentous achievements of the Nobel and Lasker Laureates of Chinese descent in literature and the sciences, we hope to inspire budding scholars to work hard and equip themselves with the necessary educational know-how to face the challenges of the 21st century. Given the space constraints, our aim in this book is to provide concise descriptions of each laureate with the hopes of piquing enough of the reader's interest to pursue more detailed research elsewhere.

Let us begin by savoring a Tang Dynasty poem:

遊子吟
作者: 孟郊

慈母手中線, 遊子身上衣;
臨行密密縫, 意恐遲遲歸。

誰言寸草心, 報得三春暉?

A Wanderer Son's Song
Author: Meng Jiao

The thread in the hands of a benevolent mother,
the resultant gown meant for a wanderer son;
sewing vigorously prior to his imminent departure,
reflecting her worries about the possibility of a belated return.

Who says that the bestowal of such tenderness to the young, deserves the reward of the sunniness of three Springs?

Literary and scientific scholars play not only an important role in bringing joy and sustenance to the mind, but also in exploring the frontiers of civilization. These scholars hail from all over the globe, and the Chinese are no exception.

Over several millennia, China has been blessed by a plethora of literary giants who have enriched us with timeless works of classic literature, including the above popular verse. Each dynasty has produced literary superstars. One reason for the abundance of exemplary scholarly giants in the past can be traced, in part, to the adoption of the Imperial Examination System. In old China, the best way to excel in life was to be an outstanding scholar and to pass with flying colors the step-wise examination series in the System. Success in these examinations could spell wealth, prosperity, desirable spouses, respect and attractive governmental offers. Indeed, the Chinese sayings that, (a) "Everything else is inferior, only scholarship is supreme (萬般皆下品, 唯有讀書高)," and, (b) "There are golden houses as well as jade-like beauties within books (書中自有黃金屋, 書中自有顏如玉)," appealed to the populace of the Middle

Kingdom. Confucius, one of China's most revered and influential philosophers, was a great proponent of education and did extol in "Lun Yu (論語)": "Is it not a great happiness if one could study often? (學而時習之，不亦說乎?)" From such a long tradition of academic pursuit deeply ingrained in the Chinese culture, it is small wonder that many Chinese are inspired to study hard!

A piece of good news centers on the fact that, in the recent past, we have witnessed the ascendancy of some outstanding Chinese language adherents and Chinese literature translators of both non-Chinese and Chinese descent, including Cecilia Lindqvist (林西莉), Michael Kahn-Ackermann, Howard Goldblatt (葛浩文), Göran Malmqvist (馬悅然), David Hawkes (霍克思), Arthur Waley (亞瑟·偉利), Franz Kuhn, Anna Holmwood, Martina U. Hasse (郝慕天), Liliane Dutrait, Noël Dutrait, Mabel Lee (陳順妍), Anna Gustafsson Chen (陳安娜) and Gilbert Chee Fun Fong (方梓勳), to name just a few. As a greater number of talented scholars come to the forefront, we expect more Chinese literary works to be included in the global canon.

With respect to scientific accomplishments, it is generally agreed that the Chinese did invent gun powder, printing, and the compass, among other significant inventions. However, economic and intellectual development stagnated in China during the Industrial Revolution due in large part to the incompetence, ignorance, corruption and hubris of the late Qing Dynasty. As a result, China became weak, laggard and poor during much of the 19th and 20th centuries in spite of its large land mass and abundant resources.[1] The blatant weaknesses exposed China to foreign invasions and ushered in an unprecedented "century of humiliation."

In terms of China's decline during the above period, however, Cardwell asserted that the old China had itself to blame:

"In brief it seems probable that imitation, adoption and adaptation are essential steps whereby the art of invention is transmitted from one culture to another. But willingness to submit to instruction, as it were, must be there in the first place. An interesting instance of what happens when the essential humility is lacking was provided by China. The Chinese produced a remarkable number of inventions and they were either willing to allow or unable to prevent the spread of these ideas to other cultures. On the other hand, the Chinese seem only rarely and in special instances to have been willing to allow the importation of foreign inventions. The consequence was that Chinese technics eventually languished."[2]

No less authorities than Fairbank *et al.* clarified the traditional China's resistance to change in the old China as follows:

"The main fact influencing China's modern transformation was that China's center of gravity lay deep within. Her long history as the ancient center of East Asian civilization had given her people an inborn sense of superiority to all outsiders. The inertia and persistence of traditional patterns and both material and intellectual self-sufficiency all made China comparatively resistant and unresponsive to the challenge of the West."[3]

In a reaction to the past, the contemporary Chinese government and culture have placed learning in high priority. Many recent Chinese strides in literature and in the sciences are being recognized globally, including the winning of a Nobel Literature Prize by Mo Yan and that of both a Nobel Medicine Prize and a Lasker-DeBakey Clinical Medical Research Award by Youyou Tu.

The list of winners of the Nobel and Lasker Awards included many Laureates of Chinese descent. In this book, we have recorded the stories of 11 Nobel Laureates, 4 Lasker Laureates (one is also a Nobel Laureate) and 1 Special Feature Wolf Laureate. Seven were Laureates in Physics; 2 in Chemistry; 2 in Literature; and 4 in Medical Sciences. A majority of these Laureates did their prize-winning work in well-developed countries such as the US and the UK. Of the 15 Laureates revered here, 4 obtained their highest degrees from the University of California at Berkeley and 3, from the University of Chicago. In addition, the remaining Laureates earned their highest degrees from the following institutions: the University of Michigan; the University of London; the Cambridge University; the Beijing Foreign Studies University; the Beijing Normal University; the Peking University School of Medicine, Beijing; the Mukden Medical College, Shenyang; and the University of Hong Kong.

All these Laureates have made significant contributions to their respective fields. This book chronicles aspects of these Laureates' lineage, upbringing, family support, education, work experiences, struggles and successes. A recurring theme in the lives of these luminaries is that success smiles on those with a strong work ethics. It is the purpose of this book to motivate people throughout the world to excel in their studies so that they can give back to their communities and the world at large! We look forward to the day when a greater number of scholars from more parts of the globe are awarded the prestigious Nobel or Lasker Prizes!

We marvel at the following Chinese saying, "There are no particular genes specifically designed to enable individuals to become generals or prime ministers; it is up to boys [and girls] to work hard to strengthen their own standings (將相本無種, 男兒當自強)."

Because our book is a collection of efforts put forth by a host of individual contributors, the length and content of each chapter may vary, but this variation has no bearing on the worth of any given Laureate. Indeed, we would like to emphasize that the Laureates in each category are of equal caliber and distinction.

We would like to thank wholeheartedly our learned and distinguished contributors for their valiant efforts to make the publication of this book possible. The overwhelming kindness and magnificent support showered on us by these individuals will always be fondly remembered and dearly treasured. We also wish to express our deepest gratitude to our publishers, Ms. Sook Cheng Lim and her colleagues at the World Scientific Publishing Company for the extraordinary job that they have done with this book.

Without the outstanding assistance of these talented professionals, this book would not have been able to see the light of day!

Finally, we would like to express our gratitude to Wikipedia from which the personal information pertaining to some of our Laureates was obtained.

References

1. Schell O, Delury J. Wealth and Power — China's Long March to the Twenty-first Century. Random House, New York, N.Y., 2013.
2. Cardwell DSL. Turning Points in Western Technology. Science History Publications, New York, N.Y., 1972, pp. 5 and 6.
3. Fairbank JK, Reischauer EO, Craig AM. East Asia — Tradition and Transformation. Revised Edition. Houghton Mifflin Company, Boston, 1989, pp. 435 and 622.

**Todd S. Ing, Keith K. Lau, Joseph M. Chan,
Hon-Lok Tang, Angela T. Hadsell and Laurence K. Chan.**

Acknowledgments

We wish to thank immensely: (a) Mr. Ho Fan for giving us permission to use his black-and-white photos depicting scenes from Hong Kong in the latter half of the 20th century, and (b) Mr. Ray Huang as well as Drs. Laurence K. Chan, Richard Yu, Shung Man Kurt Lee and Mohamed A. Rahman for their generous support of this book.

A splendid view of a Nobel Prize Award Ceremony at the Stockholm Concert Hall.
(Photo credit: TT News Agency/AFLO.)

Contents

Preface   iii

List of Contributors   xiii

A. Nobel Laureates   1

1. Chen Ning Yang, 楊振寧, 1957 Nobel Laureate in Physics   3
   Joseph J. Y. Sung

2. Tsung-Dao Lee, 李政道, 1957 Nobel Laureate in Physics   19
   Thomas T. Y. Wong, Jin Chen and Todd S. Ing

3. Samuel Chao Chung Ting, 丁肇中, 1976 Nobel Laureate in Physics   38
   Angela Y. M. Wang, Steve Siu-Man Wong and Kai-Ming Lee

4. Yuan Tseh Lee, 李遠哲, 1986 Nobel Laureate in Chemistry   56
   Yi Zhou, Jian Sun and Guanhua H. Chen

5. Steven Chu, 朱棣文, 1997 Nobel Laureate in Physics   71
   S. M. Kurt Lee, Hon-Lok Tang, Ramin Sam and Susie Q. Lew

6. Daniel Chee Tsui, 崔琦, 1998 Nobel Laureate in Physics   85
   Huai-Bin Zhuang, Xin-Cheng Xie and Fu-Chun Zhang

7. Xingjian Gao, 高行健, Literature, 2000   101
   Mabel Lee

8. Roger Yonchien Tsien, 錢永健, 2008 Nobel Laureate in Chemistry, 2004 Wolf Laureate in Medicine   116
   Keith K. Lau, Mohamed A. Rahman, Yuzhu Bian, Yun Wang and Thomas M. S. Chang

9. Charles K. Kao, 高錕, Nobel Prize in Physics, 2009 131
   Kenneth Young

10. Mo Yan, 莫言; 2012 Nobel Laureate in Literature 147
    Victoria S. Lim

11. Youyou Tu, 屠呦呦, 2011 Lasker~DeBakey Clinical Medical
    Research Award, 2015 Nobel Prize in Physiology or Medicine 165
    Ivan Barry Pless, Ray S. Huang, Xiao Qiang Ding and C. B. Lim

B. Lasker Laureates **181**

12. Choh Hao Li, 李卓皓, 1962 Albert Lasker Basic Medical Research Award ... 183
    Keith K. Lau, Cheryl D. Lau, Daniel Tak Mao Chan and Sydney Chi Wai Tang

13. Min Chiu Li, 李敏求, (1919–1980), 1972 Albert Lasker Clinical Medical
    Research Award 200
    *Steve Siu-Man Wong, Keith K. Lau, Hung-Chun Chen, Carl M. Kjellstrand,
    Antonios H. Tzamaloukas and Yuk-Lun Cheng*

14. Yuet Wai Kan, 簡悅威, 1991 Albert Lasker Clinical Medical Research Award ... 226
    *Susie Q. Lew, Hau C. Kwaan, Joseph M. Chan, Angela T. Hadsell,
    Todd S. Ing and Laurence K. Chan*

C. Special Feature, Wolf Laureate **243**

15. Chien-Shiung Wu, 吳健雄 (1912–1997), 1978 Wolf Prize Laureate
    in Physics 245
    Cheryl D. Lau, Kevin Chow, A. Ahsan Ejaz and Keith K. Lau

   Appendix A 263
   The Nobel Prize
   Susie Q. Lew and Keith K. Lau

   Appendix B 271
   The Lasker Awards
   Susie Q. Lew and Hon-Lok Tang

   Index 281

List of Contributors

Yuzhu Bian, 邊雨竹, BEng (Beijing Technology and Business University), MS (Tsinghua University), PhD (McGill University). Staff, Beijing Hejun Consulting Company, Beijing, China. Page 116.

Daniel Tak Mao Chan, 陳德茂, MBBS and MD (HKU), FHKCP, FHKAM, FRCP (London, Edinburgh and Glasgow). Chair Professor and Yu Chiu Kwong Professor of Medicine, Chief of Nephrology, Department of Medicine, HKU, Queen Mary Hospital and Tung Wah Hospital, Hong Kong, China. Page 183.

Joseph M. Chan, 陳明浩, BEng (Stanford University), MD and PhD (Columbia University). Fellow, Memorial Sloan Kettering Cancer Center; Department of Medicine, NewYork-Presbyterian/Weill Cornell Medical Center, New York, NY, USA. Page 226.

Laurence K. Chan, 陳光輝, MBBS (HKU), DPhil (Oxford University), FRCP (London and Edinburgh), FACP, FHKAM, FHKCP (Hon). Professor of Medicine, University of Colorado School of Medicine; Director, Department of Transplant Nephrology, University Hospital Health Sciences Center, Denver, Colorado, USA. Page 226.

Thomas Ming Swi Chang, 張明瑞, Order of Canada, MDCM and PhD (McGill University), FRCPC, FRS(C). Director, Artificial Cells and Organs Research Centre; Professor Emeritus of Physiology, Medicine and Biomedical Engineering, Faculty of Medicine, McGill University, Montreal, Quebec, Canada; Honorary President, International Society of Nanomedical Sciences; Honorary President, International Society for Artificial Cells, Blood Substitutes and Biotechnology; Editor-in-Chief, *Artificial Cells, Nanomedicine and Biotechnology*; Honorary Professor, Peking Union Medical College, Beijing; Honorary Professor, Blood Transfusion Institute, Chinese Academy of Medical Sciences; Honorary Professor, Nankai University, Tianjin, China; Honorary Professor, Shantou University Medical College, Shantou, China. Page 116.

Guanhua H. Chen, 陳冠華, BSc (Fudan University), PhD (California Institute of Technology). Professor and Chairperson, Department of Chemistry and Courtesy Professor, Department of Physics, HKU, Hong Kong, China. Page 56.

Hung-Chun Chen, 陳鴻鈞, MD and PhD (Kaohsiung Medical University), PhD (Juntendo University). Professor and Chief, Division of Nephrology, Department of Internal Medicine, Kaohsiung Medical University Hospital, Kaohsiung, Taiwan. President, Taiwan Society of Nephrology. President, Asian Renal Association. Page 200.

Jin Chen, 陳進, BS and ME (Shanghai Jiao Tong University), PhD (Tokyo Institute of Technology). Professor and PhD Supervisor, Shanghai Jiao Tong University; Guest Professor, Tokyo Institute of Technology; University Librarian, Shanghai Jiao Tong University, Shanghai, China. Page 19.

Yuk Lun Cheng, 鄭玉麟, MBChB (CUHK), FRCP (Edinburgh and London), FACP, FHKCP, FHKAM (Med). Chief of Service, Department of Medicine and Intensive Care Unit, Alice Ho Miu Ling Nethersole Hospital; Honorary Consultant, North District Hospital; Clinical Assistant Professor of Medicine, CUHK, Hong Kong, China. Page 200.

Kevin Chow, 曹永真. University of Illinois at Chicago science student, Chicago, Illinois, USA. Page 245.

Xiao Qiang Ding, 丁小强, MD (Shanghai Medical College, Fudan University), PhD (Fudan University). Professor and Director, Division of Nephrology, Zhongshan Hospital, Fudan University; President, Shanghai Society of Nephrology, Shanghai, China. Page 165.

A. Ahsan Ejaz, MD (Semmelweis University), FACP, FASN. Clinical Professor of Medicine, Department of Medicine, University of Florida, Gainesville, Florida, USA. Page 245.

Angela T. Hadsell, BA (Case Western Reserve University). Executive Editor for: (a) *Artificial Organs*, and (b) *Therapeutic Apheresis and Dialysis*. International Center for Artificial Organs and Transplantation. Painesville, Ohio, USA. Page 226.

Ray S. Huang, 黃紹瑜, BA (University of British Columbia), EMBA (Shanghai Jiao Tong Uinversity), JM (East China University of Political Science and Law), PhD candidate (East China University of Political Science and Law). Suntop Healthcare (Shanghai) Corp., Shanghai, China. Page 165.

Todd S. Ing, 吳兆濤, MBBS (HKU), FRCP, FRCPC, FACP. Professor Emeritus of Medicine, Stritch School of Medicine, Loyola University Chicago, Maywood, Illinois, USA; Founding President: (a) International Society for Hemodialysis, and (b) Chinese-American Society of Nephrology. Pages 19, 226.

Carl M. Kjellstrand, MD and PhD (Lund University), FRCPC, FACP. Clinical Professor of Medicine, Stritch School of Medicine, Loyola University Chicago, Maywood, Illinois,

USA; Adjunct Professor of Medicine, SUNY Downstate Medical Center, Brooklyn, New York, USA; Docent, Karolinska Institute, Stockholm, Sweden. Page 200.

Hau C. Kwaan, 關孝昌, MD (HKU), FRCP. Marjorie C. Barnett Professor in Hematology-Oncology and Professor of Medicine, Feinberg School of Medicine, Northwestern University, Chicago, Illinois, USA. Page 226.

Cheryl D. Lau, 劉舒懷, BSc (Hon), University of Toronto, Toronto, Ontario, Canada. Pages 183, 245.

Keith K. Lau, 劉廣洪, MBBS (HKU), DABP, DABP (Nephrology), MHA, FRCPCH (UK), FRCP (Canada, Edinburgh, Glasgow, Ireland and London), FAAP, FHKCPaed, FHKAM (Paed). Honorary Clinical Professor of Paediatrics, HKU, Hong Kong and Director, International Medical Center, HKU Shenzhen Hospital, Shenzhen, China. Pages 116, 183, 245.

Kai-Ming Lee, 李啓明, BSc (HKU), PhD (California Institue of Technology). Lecturer, Department of Physics, HKU, Hong Kong, China. Page 38.

Mabel Lee, 陳順妍, BA (First Class Honors) and PhD (The University of Sydney), FAHA. Adjunct Professor of Chinese Studies, The University of Sydney, Sydney, New South Wales, Australia. Page 101.

Shung Man Kurt Lee, 李尚文, MD (University of Toronto). CEO, Biotronics Kidney Center of Beaumont, Beaumont and New Century Dialysis Center, Jasper; Texas, USA. Page 71.

Susie Q. Lew, 劉瑞娟, BS (Brooklyn College), MD (SUNY Downstate Medical Center), FACP, FASN. Professor of Medicine, George Washington University School of Medicine, Washington, District of Columbia, USA. Pages 71, 226.

C.B. Lim, 林進文, MBA (Royal Brunel University, Henley Management College, London, UK); Diploma in Management Studies and Diploma in Marketing Management, Singapore Institute of Management, Singapore; Diploma in Marketing, Institute of Marketing, London, UK. CEO Advance Renal Care (Asia), Singapore. Page 165.

Victoria S. Lim, 施韫瑜, MD (Far Eastern University, Manila, The Philippines). Professor Emeritus of Medicine, University of Iowa, Iowa City, Iowa, USA. Page 147.

Ivan Barry Pless, BA, MD (University of Western Ontario), Order of Canada, DSc (Hon) (Western University), FRCPC, FRCPCH (Hon), FCAHS. Professor Emeritus of Pediatrics, Epidemiology and Biostatistics, McGill University, Montreal, Quebec, Canada. Editor Emeritus, *Injury Prevention*. Page 165.

Mohamed A. Rahman, MD (Cairo University), FACP. Nephrology Associates of Northern Illinois; Chairperson, Department of Medicine, Alexian Brothers Medical Center, Elk Grove Village, Illinois; Clinical Associate Professor of Medicine, Loyola University Chicago, Maywood, Illinois, USA. Page 116.

Ramin Sam, BS and MD (University of Illinois at Chicago). Clinical Professor of Medicine, University of California at San Francisco; Staff Physician, Priscilla Chan and Mark Zuckerberg San Francisco General Hospital and Trauma Center, San Francisco, California, USA. Page 71.

Jian Sun, 孫健, BSc, ME (Nankai University), PhD (HKU). Director of Internal Operations, Hong Kong Technology Startup Platform, Hong Kong, China. Page 56.

Joseph Jao-yiu Sung, 沈祖堯, SBS, JP, MBBS (HKU), PhD (University of Calgary), MD (CUHK), FRCP (London, Edinburgh and Glasgow), FRACP, FAGA, FACG, FHKCP, FHKAM (Med). Academician of the Chinese Academy of Engineering; Vice Chancellor/President, Mok Hing Yiu Professor of Medicine and Professor of Medicine and Therapeutics, CUHK, Hong Kong, China. Page 3.

Hon-Lok Tang, 鄧漢樂, MBBS (HKU), FRCP (Edinburgh, Glasgow and London), FHKCP, FHKAM (Med). Consultant, Department of Medicine and Geriatrics, Princess Margaret Hospital; Clinical Associate Professor (Honorary), Department of Medicine and Therapeutics, CUHK, Hong Kong, China. Page 71.

Sydney Chi Wai Tang, 鄧智偉, MBBS, MD and PhD (HKU), FRCP (London, Edinburgh and Glasgow), FACP, FHKCP, FHKAM (Med). Chair of Renal Medicine and Yu Professor in Nephrology, HKU and Queen Mary Hospital, Hong Kong, China. Page 183.

Antonios H. Tzamaloukas, MD (University of Athens), MACP. Professor Emeritus of Medicine, University of New Mexico, School of Medicine; Courtesy Staff, Raymond G. Murphy Veterans Affairs Medical Center, Albuquerque, New Mexico, USA. Page 200.

Angela Yee Moon Wang, 王依滿, MD (University of New South Wales), PhD (HKU), FRCP (London and Edinburgh), FHKAM, FHKCP. Associate Consultant, Queen Mary Hospital, Hong Kong; Honorary Associate Professor, HKU, Hong Kong, China. Page 38.

Yun Wang, 王雲, BM (Tianjin Medical University), MS (Peking Union Medical College), PhD (McGill University). Research Associate, Peking University Third Hospital, Beijing, China. Page 116.

Steve Siu-Man Wong, 黃少民, MBChB (CUHK), MRCP (UK), FRCPC, FHKCP, FHKAM (Med). Associate Consultant, Department of Medicine and Intensive Care Unit, Alice Ho Miu Ling Nethersole Hospital; Honorary Assistant Professor, CUHK, Hong Kong, China. Pages 38, 200.

Thomas T.Y. Wong, 王騰蔭, BSc (Eng) (HKU), MS and PhD (Northwestern University). Professor, Department of Electrical and Computer Engineering, Illinois Institute of Technology, Chicago, USA. Page 19.

Xin-Cheng Xie, 謝心澄, PhD (University of Maryland). Dean, School of Physics, Peking University, Beijing, China. Page 85.

Kenneth Young, 楊綱凱, BS and PhD (California Institute of Technology). Professor of Physics; Fellow and Master of CW Chu College, CUHK, Hong Kong, China. Page 132.

Fu-Chun Zhang, 張富春, BSc (Fudan University), PhD (Virginia Polytechnic and State University). Zhou Guangzhao Professor in Natural Sciences, Department of Physics, HKU, Hong Kong, China. Page 85.

Yi Zhou, 周易, BSc (Peking University), PhD student with Professor Guanghua Chen, HKU, Hong Kong, China. Page 56.

Huai-Bin Zhuang, 莊懷玢, BSc and PhD (HKU). Administrator, School of Physics, Peking University, Beijing, China. Page 85.

CUHK = The Chinese University of Hong Kong.
HKU = The University of Hong Kong.

The beautiful and majestic main building of the University of Hong Kong, Hong Kong, China. (Courtesy of the University of Hong Kong.)

"To boldly conquer frontiers that no one has even dreamed about before."

"The selection of work in which one delights and a diligent adherence to it, are main ingredients of success." Florence Bascom, quoted by Moira Davison Reynolds. In: *American Women Scientists: 23 Inspiring Biographies, 1900–2000*. McFarland & Company, Inc., Publishers, Jefferson, North Carolina, 1999, p. 142.

"Some men see things as they are and say, why? I dream things that never were and say, why not?" Robert F. Kennedy.

Katie Reilly asked Laureate Donna Strickland, a recipient of the 2018 Nobel Prize in Physics, for her research on laser beams, the following question:

"You were a graduate student when you published this research. What advice do you have for current students in the sciences?" The answer: "Do what you want to do. I believe in always going with your gut." From *TIME* (a magazine from New York, NY, USA), October 15, 2018, p. 19.

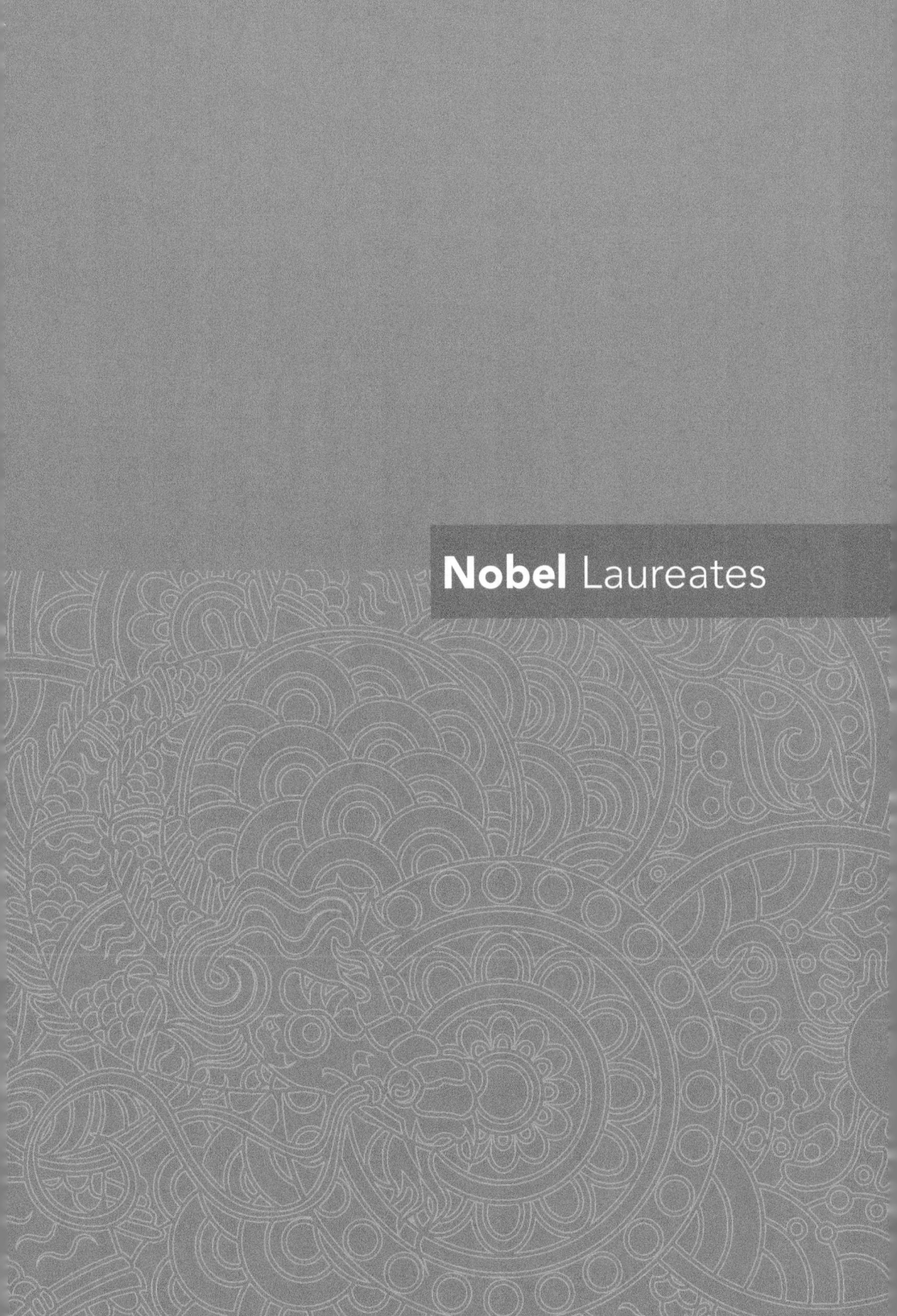

Nobel Laureates

On December 2, 1942, in this legendary and celebrated corner of the University of Chicago, the first controlled, self-sustaining nuclear chain reaction was successfully engineered by a team of the University's stellar scientists led by Professor Enrico Fermi, a superstar physicist of Italian descent. Here, Henry Moore's "Nuclear Energy" sculpture prompts us to contemplate the pros and cons of the atomic era.

Reference: How the first chain reaction changed science. The University of Chicago. (https://www.uchicago.edu/features/how_the_first_chain_reaction_changed_science/). (Photo courtesy of Dr. Laurence K. Chan.)

Chapter 1

Chen Ning Yang, 楊振寧
1957 Nobel Laureate in Physics

Joseph J. Y. Sung

Laureate Chen Ning Yang. (Photo credit: State University of New York at Stony Brook; courtesy of AIP Emilio Segrè Visual Archives, *Physics Today* collection.)

The 1957 Nobel Prize in Physics was awarded to Chen Ning Yang and Tsung-Dao Lee (李政道) based on their achievement (from Nobelprize.org): "for their penetrating investigation of the so-called parity laws which has led to important discoveries regarding the elementary particles".[1]

In 2012, I and over a hundred colleagues and students from the Chinese University of Hong Kong, celebrated the 90th birthday of Professor C. N. Yang. During the banquet, I said, "Mr. Yang, I learned about you as the first Chinese recipient of the Nobel Prize through my textbook when I was a primary school student. I would have never imagined that one day, I will be dining with you side-by-side, celebrating your birthday."

Yang and Lee's Nobel Prize discovery in 1956 centered on the parity non-conservation for weak interactions such as beta-decay — the notion that left-handed and right-handed particles do not behave in perfect symmetry as mirror images of each other. This insight demolished an unwarranted assumption held by previous physicists. This most innovative and ground-breaking idea was subsequently confirmed experimentally by Professor Chien-Shiung Wu (吳健雄) of Columbia University and scientists from the US National Bureau of Standards in 1957.

Professor Yang was the first of five children of Yang Wu-Chih (楊武之) and Lo Mong-hwa (羅孟華), and was born in Hefei (合肥市), Anhui (安徽省) on September 22, 1922, a time when China was in turmoil due to wars among warlords. His father was the first Chinese to obtain a doctorate degree from the University of Chicago. Soon after his return to China, he was recruited as a Professor of Mathematics by Xiamen University (廈門大學). Subsequently in 1929, he was also recruited as a Professor by Tsinghua University (清華大學). The young Chen Ning, therefore, lived in Xiamen, and then in Beijing, but soon moved back to Anhui to escape the fighting during the Sino-Japanese war. As the Japanese troops invaded and took over most of the cities on the East coast, the Yang family retreated to Kunming (昆明) in 1938. The young Chen Ning revealed his potential as a mathematician at the National Southwestern Associated University (西南聯合大學), a merger of Beijing University (北京大學), Tsinghua University, and Nankai University (南開大學). Inspired by the best teachers from those three major universities, Chen Ning obtained his Bachelor of Science degree at the age of 20. Then, with the aid of a fund established by the US Government after the Boxer Uprising, Chen Ning was able to further his studies at the University of Chicago and subsequently obtained his PhD, under the tutelage of Professor Edward Teller, in 1948.

Having grown up in a country in turmoil and having been educated in a university during an era of chaos, Chen Ning's passion for his motherland has never declined. In 1957, during his Nobel banquet speech, Chen Ning said, "As I stand here today and tell you about this, I am heavy with an awareness of the fact that I am in more than one sense a product of both the Chinese and Western cultures, in harmony and in conflict. I should like to say that I am as proud of my Chinese heritage and background, as I am devoted to modern science — a part of human civilization of Western origin — to which I have dedicated and I shall continue to dedicate my work." Indeed, despite spending over 50 years in the United States, successively through the University of Chicago, the Institute for Advanced Study at Princeton and the State University of New York at Stony Brook, Chinese blood is still flowing vigorously inside Chen Ning's body.

In 1971, 26 years after he had left his motherland, Chen Ning was amongst the first batch of American Chinese scientists to visit China. His visit to China has become a symbolic act, not only among Chinese scientists, but also among American scientists — China was finally opening up to the rest of the world. Not unexpectedly, he was offered the highest level of reception. He was received by the top leadership of China, including Mao, Zhou and Deng during his subsequent visits, when advising on the scientific developments of the country. He was warmly welcomed by his teachers and mentors such as Zhou Pei-Yuan (周培源), his schoolmates such as Jiaxian Deng (鄧稼先), who subsequently developed the first atomic and hydrogen bombs in China, and Huang Kun (黄昆), who was named the father of semiconductor research in China. However, what seemed to have impressed Chen Ning the most and awakened his patriotic emotion to the highest degree, is the poem of Mao's "為有犧牲多壯志, 敢教日月換新天 (Only sacrifice can foster mighty goals, brave enough to influence the sun and the moon to switch to a new sky)". These words, along with Mao's fatherly advice that China needed top-notch scientists to rebuild the country, have brought Chen Ning back to China almost every year since then. More recently, he has decided to spend more time at Tsinghua University with the goal of fostering science education.

Those who follow Chen Ning's work in physics through his myriad writings can attest to its precise elegance that was distinguished by a penchant for beauty and symmetry. To listen to his Nobel lecture, entitled "The Law of Parity Conservation and Other Symmetry Laws of Physics", is to savor and appreciate Chen Ning's feeling for the beauty of mathematical reasoning. He asserted that "when one pauses to consider the elegance and the stellar perfection of mathematical reasoning involved and contrast it with the complex and far-reaching physical consequences, a marked sense of respect for the power of the symmetry laws never fails to develop". Unlike what many people might imagine, the process of discovering scientific knowledge is not unlike that of artistic innovation. Such process is one that often is not a well-organized or carefully orchestrated intellectual endeavor. Scientific expeditions, when in the form of a part of uncharted waters or of ground-breaking moments, will require tremendous amounts of courage, a wild imagination, and an unwavering adventurous spirit. In the history of scientific exploration, the personality of the scientist may play a key role in his/her discovery. Similar to the occasion when a musician hears a few notes of music, she or he can recognize that they are coming from Mozart, Barth or Schubert. When a mathematician or physicist reads a few pages of a scientific report, he or she can immediately gauge whether the scientific background stems from Cauchy, Newton or Gauss. Consequently, it is reasonable to characterize Chen Ning as both a scientist and an artist.

Chen Ning was at a forum in Beijing University with another Nobel laureate (in Literature), Mo Yan, in 2014. They had a very open discussion about their views on many issues, including science versus art, the state of being a genius, hard work, and the "China Dream". Yang said that science is an art to predict, whereas literature is an art to fantasize; scientists never "invent", they only "discover" and continually discover. He also said everyone is born with his own talent: some of these talents are obvious, like that of Yao Ming's in playing basketball. However, talents are not always that obvious. Thus, parents and teachers need to pay attention to their charges and be vigilant. Otherwise, a lot of these talents will be wasted or be left undiscovered. Regarding the "China Dream", Yang found it difficult to express his views by using a few words. "We, the Chinese people, have gone through very tough and difficult times over the last century" asserted Yang. Ingrained deep in his mind, Yang felt that the Chinese people had been exploited or bullied by others and the reason was that we did not develop modern science. Hence, getting a Nobel Prize by a Chinese has inevitably touched the hearts of all the Chinese people. Due to the success of the economic reform, more resources are now available to help the younger generation to move forward. He deeply believes that the "China Dream" will come true one day. Yang has seen numerous Chinese youngsters who are more competent and determined than those in the West. They know exactly what they want and will work exceedingly hard to achieve their goals. Hence, he is very optimistic about the future of scientific developments in China.

As a world-renowned scientist and a physicist of the century, Yang is also a very warm and sincere person. He is a man of few words, and often, his emotions and feelings are not worn on his face. This is probably due to his upbringing as the eldest son of his family. However, deep down, he is a very warm person filled with sincerity and generosity. The co-inventor of Yang–Mills Theory, Dr. Robert Mills, described him as a brilliant star in the sky of theoretical physics. Mills also said that "Yang is an extremely generous man, a man who would share not only his office with his colleagues, but also his ideas". His schoolmates such as Huang Kun and Jiaxian Deng had similar comments about him. One of the closest allies of Chen Ning in science is Laureate Tsung-Dao Lee, the co-winner of their 1957 Nobel Prize. Between the two young physicists, they published over 30 key papers leading to the Nobel Prize.

Over the years, Yang has inspired countless young men and women, to indulge in scientific pursuits. In 1964, through the invitation of the Chinese University of

Hong Kong (CUHK), he gave an open lecture at the City Hall of Hong Kong. The lecture room was fully packed with people and there were a large number of people left waiting outside. Among the excited young people in the audience were Professor Charles Kao (2007 Nobel Laureate in Physics) and Professor Paul Chu (the Past President of the Hong Kong University of Science and Technology). In 1980, Yang accepted the appointment of Professor-at-Large at CUHK and to date, he is still mesmerizing students with his wisdom and charisma. In 1999, he decided to store his medals and hand-scripts at the CUHK's archives. Thus, this valuable collection will continue to inspire young people for generations to come. Because of his myriad exemplary contributions to CUHK, the latter decided to install a statue of Yang on its campus. Yang chose to place his statue on the roof garden of the Science Faculty that overlooks the University Mall. "I chose to stand here because I want to see the Faculty of Science continue to grow with time, and generations after generations of young scholars receiving their degrees at the congregations held in the University Mall," ventured Laureate Yang.

In my attempt to further explore the momentous contributions to science by Laureate Yang, the following citation from our university has provided a most treasured assistance: At the 53rd Congregation (1997) of the Chinese University of Hong Kong, Laureate Yang was conferred a Doctor of Science degree, *honoris causa*. The citation on Laureate Yang on that memorable occasion was distinguished by the following accolade: "Professor Yang Chen Ning has helped the rest of the scientific community to understand more about the force at work on sub-atomic particles of matter than any previous thinker in the history of science. He has shown us the universe in something far smaller than the finest dust. And his plain humanity has revealed a spirit richer than all the prizes he has won. His adventures are those of a great intellect, gleaming in the deeps of the mind."

The above CUHK citation maintained: "Earlier, in 1954, with Robert Mills, Yang Chen Ning had also formulated the theory of non-Abelian gauge fields, a unified theory by which to understand the nature of matter itself, the forces that act upon it, and the 'fearful symmetries', to use the poet William Blake's phrase, that operate to hold the formalism together. This general field-theory synthesizes at least three and possibly four of what were once thought of as four basic forces of nature. It provides a theoretical framework within which it was later shown that the seemingly separate forces are differing aspects of one force. This 'conceptual masterpiece' as it has been called, explains many features of the interaction of subatomic particles, and has redirected developments in physics especially in the last 25 years, as well as building bridges from theoretical physics to advanced mathematics." (The above citation courtesy of the Chinese University of Hong Kong.)

Personal Life[4]

In 1950, Chen Ning married Chih Li Tu. Together, they had three children, Franklin Jr., Gilbert and Eulee. Unfortunately, Chih Li passed away in 2003. Chen Ning and his second wife, Weng Fan, now reside in Beijing.

Some Other Awards[4]

- Ten Outstanding Young Americans (1957)
- Rumford Prize (1980)
- National Medal of Science (1986)
- Oskar Klein Memorial Lecture and Medal (1988)
- Benjamin Franklin Medal for Distinguished Achievement in the Sciences of the American Philosophical Society (1993)[13]
- Bower Award (1994)
- Albert Einstein Medal (1995)
- N. Bogoliubor Prize (1996)
- Lars Onsager Prize (1999)
- King Faisal International Prize (2001)

Honors

Laureate Yang has received a host of honorary degrees from various universities. In addition, he is a member of many prestigious organizations.

Editors' Note

The above apt description of Laureate Yang by the Chinese University of Hong Kong finds an ample resonance, across continents, in the citation speech given ahead of the awarding of the 2001 King Faisal International Prize in Science to our Laureate in Saudi Arabia: "Professor Yang is a renowned theoretical physicist whose research with Tsung-Dao Lee showed that the law of parity symmetry between physical phenomena occurring in right-handed and left-handed coordinate systems is violated during the decay of certain elementary particles. Prior to that, it was assumed that parity symmetry was a universal law in physics. This and other studies in particle physics earned Yang and Lee the Noble Prize in 1957. Yang's subsequent work with Robert Mills on the non-Abelian gauge theory (also known as Quantum Yang-Mills theory) laid the foundation for the unification of all interactions in nature. It is this latter work that was recognized by the

King Faisal International Prize for Science. Yang also made fundamental contributions to statistical mechanics and the theory of quantum fluid." (King Faisal Prize I Professor Chen Ning Yang, kingfaisalprize.org/professor-chen-ning-yang/.)

References and Suggested Readings

1. Chen Ning Yang — Facts — Nobelprize.org. http://www.nobelprize.org/nobel_prizes/physics/laureates/1957/yang-facts.html.
2. Chen Ning Yang — Biographical — Nobelprize.org. http://www.nobelprize.org/nobel_prizes/physics/laureates/1957/yang-bio.html.
3. Chen Ning Yang — Banquet Speech — Nobelprize.org. http://www.nobelprize.org/nobel_prizes/physics/laureates/1957/yang-speech.html.
4. Chen-Ning Yang. Wikipedia, accessed on July 12, 2016.
5. Lee T-D, Yang CN. (1956) Question of parity conservation in weak interactions. *Phys Rev* **104**: 254–8.
6. Wu CS, Ambler E, Hayward W, Hoppes DD, Hudson RP. (1957) Experimental test of parity conservation in beta decay. *Phys Rev* **105**: 1413–5.
7. Lee T-D, Yang CN. Elementary particles and weak interactions. Brookhaven National Laboratory, Associated Universities, Inc., under contract with the United States Atomic Energy Commission. Available from the Office of Technical Services, Department of Commerce, Washington 25, D.C., USA.
8. Yang CN. (2006) "Albert Einstein: Opportunity and Perception", speech at 22nd International Conference for History of Science, Beijing, 2005, *Int J Modern Phys A* **21**: 3031–8.
9. Yang CN, Mills RL. (1954) Conservation of isotopic spin and isotopic gauage invariance. *Phys Rev* **96**: 191–5.
10. Mills RL, Yang CN. (1966) Treatment of overlapping divergences in the photon self-energy function. *Prog Theor Phys Sup* **37**: 507.
11. Wu ACT, Yang CN. (2006) Evolution of the concept of the vector potential in the description of fundamental interactions. *Int J Modern Phys A* **21**: 3235–77.
12. Yang CN, Ge M-L. (2006) Bethe's hypothesis. *Int J Modern Phys B* **20**: 2223–5.
13. Brink L, Phua KK. (2016) *Proceedings of the Conference on "60 Years of Yang-Mills Gauge Field Theories. C.N. Yang's Contributions to Physics."* World Scientific Publishing Co., Singapore.
14. Chen Ning Yang. Homepage. http://insti.physics.sunysb.edu/~yang/Chen Ning Yang. Albert Einstein Professor Emeritus Nobel Laureate in Physics.
15. Chen-Ning Yang. Famous Scientists. http://www.Famous Scientists.org.

16. Nobelprize.org. Chen Ning Yang — Other Resources. Links to other sites:
    (a) Chen Ning Yang's page from C.N. Yang Institute for Theoretical Physics.
    (b) Chen Ning Yang's page from Chinese University of Hong Kong.
    (c) "Chen Ning Yang, Weak Interactions, and Parity Violation" from DOE R&D Accomplishments.
    (d) Oral History Transcript — an interview with Chen Ning Yang from the American Institute of Physics.
17. Chiang Tsai-Chien. (2013) Translated by Wong Tang-Fong. *Madame Wu Chien-Shiung: The First Lady of Physics Research*. World Scientific Publishing Co., Singapore.
18. Israel J. (1999) *Lianda — A Chinese University in War and Revolution*. Stanford University Press, Palo Alto.

Photo 1.1. Chen Ning, 10-months old, with parents, taken outside their home in Si Gu Xiang, Hefei Shi (合肥市四古巷), China in 1923. (Courtesy of Laureate Yang.)

Photo 1.2. A young Chen Ning, 3-years old, with the promise of a bright future, in 1926. (Courtesy of Laureate Yang.)

Photo 1.3. Teaching alongside Tsung-Dao Lee in 1957. (Courtesy of the Archives of the Institute for Advanced Study.)

Photo 1.4. Chen Ning and Tsung-Dao Lee at the Nobel Prize Ceremony in Stockholm in 1957. (Courtesy of World Scientific Publishing Co.)

Photo 1.5. Chen Ning and Tsung-Dao Lee receiving their Nobel Prizes in Physics from the King of Sweden, Gustaf VI Adolf, in Stockholm in 1957. (Photo credit: SCANPIX SWEDEN/Sipa USA.)

Photo 1.6. Chen Ning, Tsung-Dao Lee, Chien-Shiung Wu (front row, second from left) and fellow honorees receiving honorary doctoral degrees from Princeton University in 1978. (Courtesy of World Scientific Publishing Co.)

Photo 1.7. Chen Ning, Tsung-Dao Lee and other fellow Nobel laureates in Physics at a physics meeting. (Courtesy of Tsung-Dao (T.D.) Lee Library.)

Photo 1.8. President Ronald Reagan presenting a Medal of Science award to Chen Ning. (Photo credit: Mary Anne Fackelman-Miner, The White House, courtesy AIP Emilio Segrè Visual Archives, *Physics Today* Collection.)

Photo 1.9. Retirement ceremony for Chen Ning at the C.N. Yang Institute for Theoretical Physics, State University of New York at Stony Brook in May 1999. The eminent scientists attending the ceremony included Samuel Chao Chung Ting, Freeman Dyson, Maurice Goldhaber, Jack Steinberger, L. Cooper, J. Cronin, L. D. Fadeev, Gerardus 't Hooft, Martinus Veltman, I. M. Singer, R. Baxter, Earnest Courant and M. Rosenbluth. (Courtesy of Laureate Yang.)

Photo 1.10. Laureate Yang receiving the Doctor of Science, *honoris causa*, degree from the Chinese University of Hong Kong, along with co-honorees Professor Arthur Kwok Cheung Li, former Vice-Chancellor of the Chinese University of Hong Kong (left of photo) and Professor Ho Man Wui, Richard (right of photo). (Courtesy of the Chinese University of Hong Kong.)

Photo 1.11. With Vice-Chancellor and Professor Joseph Sung of the Chinese University of Hong Kong. (Courtesy of Vice-Chancellor Sung.)

Photo 1.12. Laureate Yang's statue is strategically positioned on the roof garden of the Faculty of Science building at the Chinese University of Hong Kong so that he could bear witness to the growth of the faculty as well as celebrate the blossoming of future scientists. (Courtesy of the Chinese University of Hong Kong.)

The gate of the National Southwestern Associated University (abbreviated as Lianda, 聯大), a university that successfully nurtured the future, first two Nobel Laureates of Chinese descent, namely, Laureates Chen Ning Yang and Tsung-Dao Lee. (Photo credit. Public domain/Wikimedia Commons.)

"When the Second Sino-Japanese War broke out in 1937, Peking University, Tsinghua University and Nankai University, merged to form Changsha Temporary University in Changsha, and later **National Southwestern Associated University** (國立西南聯合大學) in Kunming and Mengzi, in Southwest China's Yunnan Province. After the war, the universities moved back and resumed their operation." (From Wikipedia, accessed on September 9, 2016.)

Chapter 2

Tsung-Dao Lee, 李政道
1957 Nobel Laureate in Physics

Thomas T. Y. Wong, Jin Chen
and Todd S. Ing

Photo of Laureate Tsung-Dao Lee [Courtesy of Tsung-Dao (T. D.) Lee Archive Online; copyright 2012 Shanghai Jiao Tong University.]

Nobel Prize motivation: "for their penetrating investigation of the so-called parity laws which has led to important discoveries regarding the elementary particles".

Tsung-Dao Lee is a Chinese-born American physicist best known for his work on parity non-conservation in weak interactions, with which he and Chen Ning Yang (楊振寧) won the Nobel Prize in Physics in 1957. Lee has made significant contributions to a wide range of research areas, including particle physics, relativistic heavy ion physics, non-topological solitons, astronomy, statistical mechanics and condensed matter. Becoming a Nobel Laureate at the age of 31, Lee was the third youngest in history, after W. L. Bragg (at age 25 with his father W. H. Bragg in 1915) and Werner Heisenberg (at age 30 in 1932). Lee and Yang are the first Chinese Nobel laureates. Since Lee became a naturalized American citizen in 1962, he is the youngest American ever to have won a Nobel Prize.

Lee was born in Shanghai on November 24, 1926 to a scholarly family of Suzhou (蘇州) ancestry. Lee's grandfather Chong-Tan Lee (李仲覃) graduated from Bo-Shi College (博習書院) in Suzhou [became Soochow University (東吳大學) later and then Suzhou University (蘇州大學) since 1982] and was bestowed an honorary doctorate in 1920 by Randolph-Macon College of the United States. He was the first Chinese Catholic rector of St. John's Church in Suzhou. Lee's grandmother was a descendent of Ting-Shi Jiang (蔣廷錫), a renowned painter in the Qing Dynasty. His great-grandfather, grandfather and granduncle were among the founders of Soochow University. Some of these relatives were faculty members there for many years. Lee's father Tsing-Kong Lee (李駿康) was one of the first graduates in agricultural chemistry at the University of Nanking. As a chemical industrialist and merchant, he was involved in the early development of synthetic fertilizers in China. Although he was busy as a successful industrialist, Tsing-Kong Lee gave much attention to the education of his children. Each child received private tutoring in fundamental subjects while being immersed in the atmosphere of learning fostered in the Lee family. Every sibling of Lee is a university graduate.

In childhood days, Lee was an avid reader, covering a wide range of subjects. His mother, Ming-Chang Chang (張明璋), who studied at Chi-Ming High School (啟明中學) in Shanghai, was a highly intelligent woman. She often took Lee to browse in bookstores such as Commercial Press, Sinhua, and Chi-Ming in Shanghai and would readily purchase for him the ones in which he was interested. *The Adventures of Tom Sawyer* by Mark Twain and *The Expanding Universe: Astronomy's "Great Debate", 1900–1931*, by Arthur Eddington were two books among his favorites.

Lee attended the high school affiliated with Soochow University in Shanghai, and transferred to Jiangxi Joint High School (江西贛州聯合中學) in Gongzhou (贛州),

Jiangxi province (江西省), after the eruption of the Second Sino-Japanese war. This was where Lee's interest in physics developed, owing much to a nearby library which had a vast collection in natural sciences. However, his education was interrupted by the war and he did not receive his high-school diploma. Enduring enormous hardship, Lee travelled for two months by way of Guangdong and Guangxi to attend the joint university entrance examination in Guiyang in 1943. Since he had not officially graduated from high school, his candidacy was granted by equivalent credentials. His outstanding examination scores qualified him for acceptance to the electrical engineering program at Zhejiang University (浙江大學), which had recently relocated to Guiyang. His talent and interest in physics were quickly recognized by several physics professors, including Hsin Pei Soh (束星北) and Ganchang Wang (王淦昌), who guided his entrance to the world of physics and helped him arrange for his transfer to the physics department. In 1944, the conditions in war-torn Guiyang threatened the academic vitality of Zhejiang University. After travelling to Chungking to meet his mother and brothers (on the way he was hospitalized for severe injury caused by an automobile accident), Lee transferred to the National Southwestern Associated University (NSWAU) in Kunming (昆明).

Lee arrived at Kunming in the middle of the school year. With a letter of introduction written by a friend of his aunt to Ta-You Wu (吳大猷), renowned physics professor at NSWAU, Lee was recommended by Wu to audit second-year courses and took examinations, which he breezed through with flying colors. Lee often approached Wu for advanced reading material and exercises, which he delved into with much compassion. Eventually, Lee became a frequent visitor to Wu's home where he also helped with household work and the treatment of Wu's back pain. After the end of the war, a group of six students were selected to receive the Chinese government fellowships to pursue graduate studies in the United States. Lee was one of them, upon the recommendations of Wu and colleagues Loo-Keng Hua (華羅庚) and Chao-Lun Tseng (曾昭掄), and the endorsement from Chi-Sun Yeh (葉企孫), a physics professor and the dean of the College of Science at NSWAU.

Although there was no doubt from Wu and his colleagues that Lee's capability was commensurate with that of a high-caliber graduate student, his official record indicated that he had completed only second year undergraduate courses. Arriving in the United States in 1946, Lee had to overcome the hurdle of gaining admission to a graduate program. Fortunately, the University of Chicago had the provision to grant admission based on equivalent qualifications, but not without tacit negotiations from Lee. His talents were quickly recognized by the Chicago faculty after his enrollment. Among them was Edward Teller, in whose course on quantum mechanics, Lee provided a lucid solution to a problem employing fractional calculus, which deeply impressed Teller. When Enrico Fermi heard about Lee's ability, he began to pay attention to Lee and invited him to attend his special seminars. After passing the qualifying examination for the doctoral program, Lee became a student of Fermi's. While Lee has been a theoretician in his entire career, he is also savvy as a practitioner. When he was studying the temperature distribution of the sun, a need arose to solve a system of coupled differential equations. In the days before electronic computing, solving these equations and obtaining numerical solutions were a cumbersome process. Lee and Fermi constructed a huge slide rule that was customized to obtain the numerical solutions to the system of equations.

Lee first met Chen Ning Yang at NSWAU and reunited with him in Chicago. Yang was older than Lee and had arrived in Chicago earlier, receiving his doctorate in 1946, with a thesis written under the supervision of Teller. Lee's first scientific paper was co-authored with M. Rosenbluth and Yang, on the subject of particle interactions. This paper paved the way for the development of the theory of weak interaction and its application to processes other than beta decay. Rosenbluth was a student of Teller but shared an office with Lee. Together with J. Steinberger, also a student in Fermi's group, Lee was investigating the decay and capture of mesotron. Yang often had dinner with Lee on weekends, during which physics research was discussed. Yang was very interested in the theoretical work on mesotron and joined in the effort, leading to the publication of the paper in 1949 in the scientific journal *Physical Review*. That publication much impressed Fermi, who devoted a substantial portion of his book *Elementary Particles* to elaborate on the results obtained by Lee, Rosenbluth and Yang.

Lee continued to pursue research on astrophysics, obtaining a new value for the Chandrasekhar Limit as 1.44 times the sun's mass for the white dwarfs. He received his doctorate in June 1950 with a dissertation on the hydrogen content of white dwarf stars. Upon completion of his dissertation in 1949, Lee joined S. Chandrasekhar (a University of Chicago faculty member who received a Nobel Prize in Physics in 1983) at the Yerkes Observatory in Wisconsin for eight months before joining the University of California in Berkeley as a lecturer. In 1951, Lee joined the Institute of Advanced Studies where he and Chen Ning Yang made important contributions to statistical mechanics. Lee also spent the summer of 1952 at the University of Illinois to interact with scientists to conduct research on condensed matter physics.

Lee joined the faculty of Columbia University in 1953 and became a full professor at the age of 29 in 1956. In 1954, he published a solvable model of quantum field theory, known as the Lee model, which provided insight into the development in quantum chromodynamics. His focus turned to particle physics, especially the developing puzzle of K meson decays. It was observed in all strong interactions and electromagnetic interactions that reflection symmetry holds. In simple terms, a physical process and its mirror image are expected to uphold the laws of physics. Known as the conservation of parity, this "principle" was taken to be valid for an extended period of time.

The tau (τ)-theta (θ) puzzle related to subatomic particles was hotly debated in the early 1950s. Based on one set of criteria, that of mass and lifetime, two elementary particles, τ and θ, appeared to be the same, whereas on another set of criteria, they appeared to have different spin and intrinsic parity. Parity is an important characteristic in quantum mechanics because the wave functions that represent particles can behave in different ways upon transformation of the coordinate system that describes them.

Subatomic particles have various properties and are affected by certain forces that exhibit symmetry. An important property that gives rise to a conservation law is parity.

Until 1956 it was assumed that, when an isolated system of fundamental particles interacts, the overall parity remains the same or is conserved. This conservation of parity implied that, for fundamental physical interactions, it is impossible to distinguish right from left and clockwise from counterclockwise. The laws of physics, it was thought, were indifferent to mirror reflection and could never predict a change in parity of a system.

In physics, all known forces in the universe can be grouped into four basic types: The strong force responsible for binding of nuclei; the electromagnetic force between all particles which have electric charge; the weak force that is responsible for nuclear beta decay and other similar decay processes involving fundamental particles; and the gravitational force that holds us onto the Earth. Lee and Yang examined the evidence for parity conservation and found, to their surprise, that although there was strong evidence that parity was conserved in the strong (nuclear) and electromagnetic interactions, there was, in fact, no supporting evidence that it was conserved in the weak interaction. They published two papers in 1956 and 1957 in *Physical Review* that made a systematic study of possible P, T, C and CP violations in weak interactions (P refers to transformation by mirror reflection, T refers to time reversal, C refers to replacement of particles by antiparticles, and CP refers to a combination with C followed by P transformation) and provided descriptions for experiments to verify the predictions of parity non-conservation. With the dedicated effort of Chien-Shiung Wu (吳健雄) (please see Chapter 15), an experimentalist of the physics faculty at Columbia, and her collaborators at the National Bureau of Standards in Washington, D.C., the suggestion of parity non-conservation in beta decay from radioactive cobalt was confirmed. Results of this experiment were published in *Physical Review* in 1957. Leon Lederman and his colleagues designed an alternative experiment to disprove the law of parity at the Nevis Laboratory of Columbia University. Since Wu and her colleagues discovered the non-parity phenomenon first, their article was placed ahead of Lederman's in the 1957 issue of *Physical Review*. Thereafter, parity non-conservation was confirmed by numerous other laboratories. In October 1957, Lee and Yang were awarded the Nobel Prize in Physics, "for their penetrating investigation of the so-called parity laws, which has led to important discoveries regarding the elemental particles".

Lee and Yang had neighboring offices when Lee was on leave from Columbia at the Institute of Advanced Studies in Princeton in 1960. Their working conversations were often overheard by physicist Jeremy Bernstein, who described them as "going at any job with tremendous gusto, and usually at full volume. They take great pleasure in racing each other in calculations, and as they are extremely fast thinkers, watching them or listening to them at work can be both an exhilarating and somewhat exhausting experience."

Around 1959, Lee, Yang, and collaborators initiated the field of high energy neutrino physics. Subsequently, a general method for treating divergences connected with particles of zero rest mass was introduced. Lee and Chien-Shiung Wu published a series of papers in two groups on the subject of weak interactions in 1965–1966. His publication of several papers on a new form of matter in high density in 1974–1975 paved the way for the modern field of relativistic heavy ion collider (RHIC) physics. Lee and collaborators commenced work to establish the field of non-topological solitons, which led to his work on soliton stars and black holes throughout the 1980s and 1990s.

Besides research and teaching, Lee has devoted his time to administrative and advisory tasks. From 1997 to 2003 Lee was director of the RIKEN-BNL [The Institute of Physical and Chemical Research (of Japan) and the Brookhaven National Laboratory] Research Center (currently director emeritus). Joining efforts from other researchers from Columbia, the Center completed a 1 teraflops supercomputer QCDSP for lattice QCD in 1998 and a 10 teraflops QCDOC machine in 2001. Most recently, Lee and R. Friedberg have developed a new method to solve the Schrödinger Equation, which provided convergent iterative solutions for the long-standing quantum degenerate double-wall potential and other instanton problems. They have also done work on the neutrino mapping matrix.

An area that Lee has devoted much attention to throughout his career is statistical mechanics. From the early days of his career, Lee has applied knowledge in statistical mechanics to research problems in physics, drawing important conclusions while contributing to the field of statistical physics itself. His doctoral thesis, written under the tutelage of E. Fermi, addressed the hydrogen content of white dwarf stars, for which statistical techniques were employed. In the year 1952, Lee and Yang, both at the Institute of Advanced Studies in Princeton at that time, jointly authored two classic papers in *Physical Review* that put the equations of states and discontinuities in thermodynamic functions associated with phase transitions on a rigorous mathematical foundation. With the proofs of several encompassing theorems, they removed certain inadequacies in the theory of phase transition at that time. Besides the lucid account of concepts in physics, the work described in those papers was a *tour de force* in mathematical physics. The intellectual capacity of the authors was certainly crucial in achieving the outcomes of the research. But their education background also had a significant bearing in fostering the elegant work. It would therefore be of interest to trace the path pursued by Lee in his acquisition of knowledge in statistical mechanics.

Lee began studying physics in Zhejiang University, with prominent professors including Hsin Pei Soh and Ganchang Wang. Soon after joining the National Southwestern Associated University in Kunming, Lee had Jwu Shi Wang (王竹溪) as instructor for courses in thermodynamics and statistical mechanics, giving him a solid foundation in this area. Wang was a prolific scholar in physics, who was the author of China's benchmark textbooks on these two subjects. Wang studied statistical physics at Cambridge University with R. H. Fowler, an authority in the field. During the days in Cambridge, Wang was a close friend of P. A. M. Dirac, also a student of Fowler. After returning to teach in Tsinghua University, Wang put in much effort to promote the physical principles that belie the mathematical formulism and notations introduced by Dirac to quantum mechanics. In Lee's student days at Zhejiang and Tsinghua, he had a habit of solving all problems given as exercises in textbooks, which provided him with special skills and insight to tackle mathematical challenges in scientific research. It would not be illogical to suggest that Lee's rigor in mathematical methods was instilled and fortified by Wang, who was the co-author of a celebrated text on special functions, in addition to several books in physics mentioned earlier and a conventional Chinese dictionary in traditional characters.

The educational experience Lee had while studying at the University of Chicago provided him with the opportunity of working with the pioneers in statistical mechanics of the time. His mentor, E. Fermi, shared with Dirac the namesake of Fermi-Dirac statistics, which are obeyed by particles that subscribe to Pauli's exclusion principle, such as the electron. Lee took courses in statistical mechanics jointly taught by J. Mayer and M. Goeppert-Mayer (Nobel laureate in Physics, 1963). They provided Lee with the unique insight into the success and limitation of Mayer's theory of phase transition, paving the way for the work that he later carried out with Yang on the Ising model and phase transition.

In the summers of 1952 and 1953, Lee took up the invitation of J. Bardeen to visit the campus of the University of Illinois at Urbana-Champaign to collaborate with F. Low and D. Pines to work on research in condensed matter physics. Although the work was not concentrated in statistical mechanics, concepts in statistical mechanics and techniques of analysis often used in the field were brought in for studying the properties of the polaron. The paper on the subject of electron motion in polar crystals that they published in *Physical Review* and a subsequent lecture delivered by Lee at a conference in 1957, provided the inspiration for a key step in the development of the BCS theory for superconductivity (BCS stands for Bardeen, Cooper and Schrieffer). For this work on superconductivity, Bardeen shared his second Nobel Prize with Cooper and Schrieffer in 1972.

The period of 1957 to 1960 saw highly significant results coming out of the collaboration of Lee, Yang, and Kerson Huang (黄克孫) on quantum statistical mechanics. A general framework for treating many-body systems was established, along with an incisive investigation of a Bose system of hard spheres, which was readily applied to explain superfluidity, phonon and sound velocities, and other peculiar properties exhibited by liquid helium.

There was a period of time when Lee directed his major research effort to other branches of physics but he stayed abreast of the development in statistical mechanics. He later returned to conduct research in the field and applied the techniques to topics at the forefront of physics, an example being the work he published with R. Friedberg and H. C. Ren on fullerenes (C_{60}) in 1992. When Lee was invited to deliver lectures to a wide audience in Beijing in 1979, statistical mechanics was one of the three subjects that he chose to speak on.

Beginning with his return visit to China in September 1972, Lee has made numerous contributions to strengthen basic science education and physics research in China. He advocated the building of the electron-positron collider in China (commenced in 1984 and completed in 1988), promoted the collaboration of US/China in high energy physics, and facilitated the first meeting of the US/PRC Joint Committee on High Energy Physics at the SLAC National Accelerator Laboratory (SLAC is the abbreviation for the Stanford Linear Accelerator Center) in 1979. A historic moment was reached in 1984 when Lee invited Jun Shan Shen (沈君山), Taiwan Physical Society President, Guangzhao Zhou (周光召), Chinese Academy of Sciences Vice President and Ta-You Wu to convene a telephone conference hosted by Lee from his office at Columbia University.

In view of the lack of opportunities for physics students to take part in GRE (Graduate Record Examinations) and TOEFL (Test of English as a Foreign Language), scores of which are required to apply for graduate programs in the universities in the United States, Lee proposed in May 1979 the China-US Physics Examination Program (CUSPEA) to enable Chinese students to directly take qualifying examinations given by professors at US universities, initially with five students applying to Columbia, and quickly broadening to 76 US universities. This was accomplished with the untiring effort of Lee to coordinate with the administration offices of institutions in China and in the United States. Much of his time during the inaugural years of the program was devoted to proposal and letter writing, telephone calls, preparation of official documents and promoting the program to colleagues in physics departments at universities in both countries. Between 1979 and 1989, a total of 915 Chinese graduate students have received placements with financial support for graduate study at US universities.

A proposal was made by Lee in 1984 to establish postdoctoral mobile stations to enable returning doctoral graduates to continue their research in China. The program was launched in 1985 with Lee being named the National Postdoctoral CMC and Honorary Chairman of the Chinese Postdoctoral Science Foundation. (Editors'

Note: CMC Stands for Certification Management Committee). Up to 2003, a total of 947 stations have been built in universities and research institutes, and 670 in industry. In 1992, the Tsung-Dao Lee Physics Scholarship was established at Fudan University. In 1998, the "Hui-Chun Chin and Tsung-Dao Lee Endowment in Memory of Jeannette Lee" [Jeannette Lee was Hui-Chun's English name] (秦惠䇹一李政道中國大學生見習基金) was established, with donation from the couple's private savings, to support students in non-science majors to gain exposure to science by internship during holidays and after-school hours.

Visualizing that science and art were intertwined in their development in early civilizations, Lee believes that the two fields may fortify each other when properly facilitated to enable the interactions between scientists and artists. Advancements in science and arts rely on human intelligence and creativity, as he often stresses, while an important goal common to both is to seek the truth and eternal values in human endeavors. Lee invited renowned painters, among them Zuoren Wu (吳作人) and Shufang Xiao (肖淑芳), professors at the China Central Academy of Fine Arts, to produce theme paintings for scientific conferences in China, which developed into a forceful trend enabling cross fertilization between the two fields. Lee and many artists became close friends as he promoted the works of the artists from the perspective of science. The collection of Lee's own work in art domains includes paintings, poems and calligraphy.

In 2011, nearly six decades after joining Columbia, Lee retired as University Professor. At a reception in his honor at the President's House on September 13, Columbia President Lee Bollinger singled out Lee's staggering, scholarly contributions to the field of physics. "T. D. Lee has succeeded into his 84th year in continuing to push the boundaries of the study of particle physics," said Bollinger, "and in every important way, his unsurpassed career represents all that is best about Columbia University and about the academy."

In appreciation of his city of birth, Shanghai, Lee donated his Nobel Prize medal and diploma along with a collection of his lifelong works of 70,000 items to a library named after him. The opening ceremony of this library, the T. D. Lee Library, at the campus of Shanghai Jiao Tong University was held on December 28, 2014.

"Cultivating scientific talent should not be limited to teaching in class," Lee said in a video message at the opening ceremony of the Library. "My purpose in donating all my scientific archives and manuscripts to Shanghai Jiao Tong University is to benefit later generations. I hope the Tsung-Dao Lee Library may inspire young students to climb to the heights of science," he added.

Attending the ceremony was James Lee, son of Lee and Hui-Chun, and Dean and Chair Professor of the School of Humanities and Social Sciences, Hong Kong University of Science and Technology. The library, which has been nicknamed "Nobel Hill", with photographs, academic manuscripts and valued items donated by Laureate Lee, is open to the public. These include more than 3,400 papers, 5,700 research manuscripts, 40,000 letters, scientific and artistic works, awards and books. "Visitors can see life-long scientific research of my father and the modern history of scientific development in China," James said.

On November 28, 2016, the inauguration of the T. D. Lee Institute, in recognition of T. D.'s contributions in science, was held in the T. D. Lee Library on the Campus of Shanghai Jiao Tong University. This institute was created as a platform, similar to the Neils Bohr Institute of the University of Copenhagen and Institute of Advanced Study at Princeton, for scientists from all corners of the globe to collectively unlock some of the deepest mysteries of our universe.

Professor Lee has received numerous awards and honors from around the world. His recognition even extends beyond this world, for in 1997 Small Planet 3443 was named as the "3443 Leetsungdao" in his honor.

References

1. Tsung-Dao Lee — Facts — Nobelprize.org.
2. T. D. Lee, C. N. Yang. Question of Parity Conservation in Weak Interactions. *Phys Rev* **104**, 254 (1956).
3. Tsung-Dao Lee. Wikipedia, accessed July 25, 2017.
4. C. S. Wu, Adventures in Experimental Physics (Gamma volume), ed. B. Maglich, Princeton, N.J., World Science Communications, 1973, p. 102.
5. Leon Lederman, The God Particle. Houghton Mifflin Co., Boston-New York (1993).
6. C. S. Wu, E. Ambler, R. W. Hayward, D. D. Hoppes, R. P. Hudson. Experimental test of Parity Conservation in Beta Decay. *Phys Rev* **105**, 1413 (1957).
7. R. L. Garwin, L. M. Lederman, M. Weinrich. Observations of the Failure of Conservation of Parity and Charge Conservation in Meson Decays: the Magnetic Moment of a Free Muon. *Phys Rev* **105**, 1415 (1957).
8. Biography of Tsung-Dao Lee, http://en.wikipedia.org/wiki/Tsung-Dao_Lee.
9. 季承, <<李政道傳>>, 國際文化出版公司 (2010). Biography of Tsung-Dao Lee. An account of Tsung-Dao Lee's overseas education.
10. 陳典松, 李政道出國留學紀事, 《傳記文學》第 292 期, 83–91 頁 (2014). A description of the overseas studies of T. D. Lee's.
11. Biography of Jwu Shi Wang, 王竹溪生平. http://baike.baidu.com/item/%E7%8E%8B%E7%AB%B9%E6%BA%AA
12. Emilio Segrè. Enrico Fermi, Physicist. University of Chicago Press, Chicago and London (1970).
13. T. D. Lee, M. Rosenbluth, C. N. Yang. Interaction of Mesons with Nucleons and Light Particles. *Phys Rev* **75**, 905 (1949).

14. C. N. Yang, T. D. Lee. Statistical Theory of Equations of State and Phase Transitions. I. Theory of Condensation. *Phys Rev* **87**, 404–9 (1952).
15. T. D. Lee, C. N. Yang. Statistical Theory of Equations of State and Phase Transitions. II. Lattice Gas and Ising Model. *Phys Rev* **87**, 410–9 (1952).
16. Nobel Lectures, Physics: 1942–1962. Elsevier Publishing Company, Amsterdam, 1964.
17. J. Bernstein, Profiles: A Question of Parity, *The New Yorker Magazine*, 38: May 2, 1962.
18. J. Sucher, Columbia Physics in the Fifties: Untold Tales. *Mathematical Physics and Quantum Field Theory* (published July 12, 2000), pp. 197–206.
19. Lee, Tsung-Dao. *The Asian Encyclopedia*. Editor: Franklin Ng. Publisher: Marshall Cavendish Corp, 1995, pp. 985–6.

Photo 2.1. T. D. Lee (first from left) with mother and other members of family. (Courtesy of Tsung-Dao (T.D.) Lee Library.)

Photo 2.2. T. D. Lee at 14 years of age. (Courtesy of Tsung-Dao (T.D.) Lee Library.)

Photo 2.3. T.D. Lee at 30 years of age. A short time later, he won the Nobel Prize. (Photo credit: Science/AFLO.)

Photo 2.4. T. D. Lee and C.N. Yang. (Courtesy of the AIP Emilio Segrè Visual Archives.)

Photo 2.5. T. D. Lee and wife, Hui-Chun Chin. (Courtesy of Tsung-Dao (T.D.) Lee Library.)

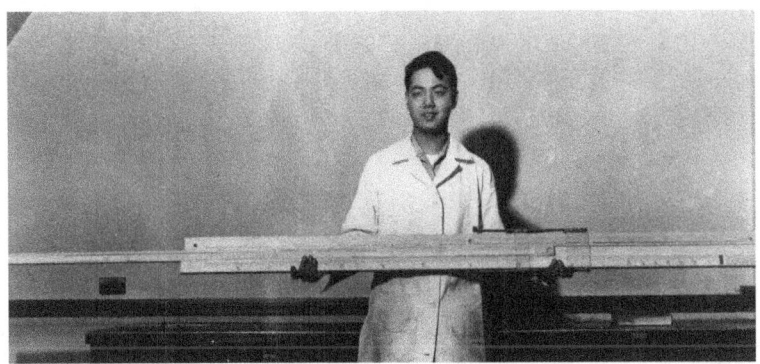

Photo 2.6. With a giant slide rule constructed with Enrico Fermi. (Courtesy of Tsung-Dao (T.D.) Lee Library.)

Photo 2.7. T. D. Lee receiving the Nobel Prize from King Carl XVI Gustaf of Sweden in 1957. (Courtesy of Tsung Dao (T.D.) Lee Library.)

Photo 2.8. T. D. Lee receiving the Nobel Prize in Physics at Stockholm. Photo of Nobel Prize winners in 1957: From left C. N. Yang (Physics), D. Bovet (Medicine), T. D. Lee (Physics), A. Todd (Chemistry) and A. Camus (Literature). (Photo credit: SCANPIX SWEDEN/Sipa USA.)

Photo 2.9. With Ta-You Wu (sitting). (Courtesy of Tsung-Dao (T.D.) Lee Library.)

Photo 2.10. T. D. Lee receiving a Doctor of Laws, *honoris causa*, degree from The Chinese University of Hong Kong during the latter's 11th Congregation in 1970. Photo taken with (from left) Ieoh Ming Pei (I.M. Peh), S. S. Gordon, Choh-Ming Li, David Trench, Tsung-Dao Lee, and Choh Hao Li. (Courtesy of the Chinese University of Hong Kong.)

Photo 2.11. At the celebration of T. D. Lee's 60th birthday (from left) T. D. Lee, Chien-Shiung Wu (吴健雄), Luke Chia-Liu Yuan (袁家骝), Ta-You Wu (吴大猷), Samuel C. C. Ting (丁肇中). (Courtesy of Tsung-Dao (T. D.) Lee Library.)

李政道夫妇与吴作人夫妇合影
Professor and Mrs T. D. Lee and Professor and Mrs Wu Zuoren

Photo 2.12. T. D. Lee and wife with artists Zuoren Wu (吴作人) (far right) and Shufang Xiao (肖淑芳, Mrs. Wu) (far left). (Courtesy of Tsung-Dao (T. D.) Lee Library.)

孟子見梁惠王. 王曰: "叟!不遠千里而來, 亦將有以利吾國乎?" 孟子對曰: "王!何必曰利? 亦有仁義而已矣."

Mencius visited King Hui of Liang. The King said: "Old fellow! You went through the trouble of traveling one thousand miles to come here, do you have a plan to bring profit to my country?" Mencius replied: "King! What is the point of mentioning the word profit? All that matters, after all, is that there should be benevolence and righteousness."

The above translation is partially derived from Sir David Todd's address to the graduates of the Faculty of Medicine, University of Hong Kong, January 15, 2011.

Reference: Medicine, a Noble Profession, by David Todd. Hektoen International, volume 8, issue 3, Summer 2016 Special Issue.

Chapter 3

Samuel Chao Chung Ting, 丁肇中
1976 Nobel Laureate in Physics

Angela Y. M. Wang, Steve Siu-Man Wong and Kai-Ming Lee

Laureate Samuel Chao Chung Ting. (Photo credit: Photo by Calvin Campbell, Massachusetts Institute of Technology, courtesy of AIP Emilio Segrè Visual Archives, Physics Today Collection, W. F. Meggers Gallery of Nobel Laureates Collection, and courtesy of MIT Museum.)

Discovery of the J/Psi Particle

By about 1970, a large number of elementary particles had been identified. Most of these comprised two or three quarks. Using high-energy particle collisions and independently of each another, in 1974 both Samuel Ting and Burton Richter, discovered a new heavy particle, known as J/psi, proving experimentally the existence of a fourth quark, later named "charm".[1]

Dr. Samuel Chao Chung Ting is an American physicist who shared the 1976 Nobel Prize Award for Physics with Burton Richter from the Stanford Linear Accelerator Center for "their pioneering work in discovering a heavy elementary particle of a new kind" called J/ψ meson nuclear particle in 1974. Hence, Ting and Richter proved the existence of a fourth quark, later named "charm". At the time of receiving the Nobel Prize, Dr. Ting was only 40-years old and was working at the Massachusetts Institute of Technology (MIT), where Dr. Ting headed a research team that explored new areas of high-energy particle physics. Dr. Samuel Ting and Dr. Burton Richter received the Nobel Prize award in less than two years after their dual discoveries, which, in Nobel Prize history, is one of the shortest time frames between a Laureate's discovery and recognition.[1-4]

Samuel was born on January 27, 1936, in Ann Arbor, Michigan, USA. He was the eldest of three children of Kuan Hai Ting (丁觀海), an engineering professor, and Jeanne Tsun-Ying Wang (王雋英), a psychology professor. Ting's parents met at the University of Michigan and married while they were graduate students. His father's family was from Rizhao County (日照縣) in the Shandong province (山東省) of China. Samuel was born prematurely while his parents were visiting the United States and, as a result, he accidentally became an American citizen. Soon after his birth, Samuel returned to China with his parents. Due to wartime conditions, Samuel's education was disrupted and he did not receive a traditional education until he was 12-years old. Nevertheless, Samuel had the opportunity to meet many accomplished scholars during his childhood since his parents were always associated with universities. Perhaps due to this early exposure to and influence from intellectuals, Samuel yearned to be associated with university life from a young age.[5-7]

After the war, Samuel's parents became professors at the National Taiwan University in Taipei, Taiwan. Beginning in 1948, Samuel attended the prestigious Provincial Chien-Kuo High School (建國中學, now called Municipal Taipei Chien-Kuo Senior High School) in Taipei. After high school, he spent one year at the National Cheng Kung University, Tainan City, Taiwan.[5] Since both his parents were extremely occupied with work, Samuel was brought up mostly by his maternal grandmother, from whom he learned many stories about the difficult lives of Samuel's mother and maternal grandmother during the wartime and about the efforts that the maternal grandmother made to provide her daughter with a good education. Both his mother and maternal grandmother were daring, authentic and determined people, and both left an unforgettable impression on Samuel.

When Samuel was 20-years old, he decided to return to the United States in pursuit of a good education. However, Samuel knew very little English at that time and was unprepared for the high cost of living in the United States. He arrived at Detroit airport on September 6, 1956, with only $100 USD. He did not know anyone, and communication was tremendously difficult. Samuel relied on scholarships to pay for his education and was motivated to work very hard. Despite the hardship, he managed to obtain his degrees in both mathematics and physics from the University of Michigan in only three years instead of the typical four, and completed his PhD degree in physics in 1962. He then relocated to work at the European Organization for Nuclear Research (CERN) as a Ford Foundation Fellow. There, he worked on the Proton Synchrotron with Giuseppe Cocconi, who impressed Samuel in the way that he always had a simple way of viewing a complicated situation and managed to handle experiments with meticulous care.

In the spring of 1965, Dr. Ting returned to the United States and taught at the Physics Department of Columbia University, which he discovered to be a most stimulating place. He interacted with other distinguished scholars, each of whom had his or her own individual style in physics. During his second year at Columbia, a major scientific event exploded in the news: at the Cambridge Electron Accelerator, an experiment had been performed on the electron-positron pair production by photon collision using a nuclear target, and the results seemed to violate quantum electrodynamics. Upon hearing such a groundbreaking news, Samuel decided to duplicate the experiment. He contacted G. Weber and W. Jentschke of the Deutsches Elektronen Synchrotron (DESY) about the possibility of doing a pair production experiment at Hamburg. Weber and Jentschke were very enthusiastic and encouraged him to start right away. In March 1966, he went on leave from Columbia University to perform this experiment in Hamburg. Since then, he has devoted all his efforts to the examination of quantum electrodynamics, the physics of electron pairs, the production and decay of photon-like particles, and to the search for new particles which will decay to electron pairs.

In 1969, he joined the Physics Department of the Massachusetts Institute of Technology. In order to search for new particles of a higher mass, he recruited his team to work in the United States in 1971 and started the experiments at the Brookhaven National Laboratory. Samuel and his team studied the collisions of particles accelerating at both very high speed and energy, meaning more than 10 billion times the typical energy of chemical reactions. Electrons are usually created from collisions. So when the energy of collisions reached threshold value that is equivalent to roughly three times the mass of a proton, then electrons may be produced and new particles may be created. The higher the energy of the colliding particles, the smaller and the heavier the created constituents will be. In other words, the fundamental idea underlying particle accelerator experiments is that only extremely high energy can give rise to heavy particles because energy and mass are related by the famous formula, $E = mc^2$, developed by Albert Einstein.

In the fall of 1974, Samuel's group discovered a new elementary particle that is 3.3 times heavier than a proton and has a much longer life than expected for particles within this mass range (here, lifespan is measured in very small fractions of a second). However, when he attended a scientific conference at the Stanford Linear Accelerator Center (SLAC) in California, there were rumors that SLAC had discovered the same particle. Samuel knew he and his team could not wait any longer. After discussion with SLAC, both teams agreed to announce the discovery together on November 11, 1974. They decided to name this new heavy elementary particle "J/Psi". The name was inspired by the Chinese character of the surname of Ting, namely, 丁.

Since this discovery, a host of new particles has been discovered.

(Please note that references 9–16 refer to the following three paragraphs).

In 1977, Dr. Ting was appointed as the first Professor of Physics at the Thomas Dudley Cabot Institute of MIT. In 1995, not long after the cancellation of the Superconducting Super Collider Project, which had severely reduced the possibilities for high-energy physics experiments on Earth, Samuel turned to outer space. Driven by the thought that a discovery of a single atomic nucleus heavier than anti-helium could signify the existence of an anti-star or perhaps even a whole anti-galaxy somewhere in the realm of the universe, Samuel said, "If you don't do the measurement, you will never know." Thus, Samuel proposed to Dan Goldin, who was the administrator of the National Aeronautics and Space Administration (NASA) at that time, that he could make a measurement using a space-based cosmic ray detector called the Alpha Magnetic Spectrometer (AMS), which is designed to sift high-energy particles, meaning cosmic rays flying through space. A prototype, the *AMS-01*, was flown in and tested on Space Shuttle mission STS-91 in 1998. The main mission, the *AMS-02*, was then planned for launch by the Shuttle and for mounting on the International Space Station. This extraordinary project aimed to advance the current knowledge of the universe, as there were still many questions unanswered. Functioning as the "Hubble Space Telescope" of cosmic rays, AMS-02 was designed to measure these rays with the utmost accuracy and to study their characteristics directly in space in order to look for convincing evidence of the existence of antimatter and dark matter. This approach is much needed in our understanding of the Universe's origin and evolution.

Unfortunately, after the 2003 Space Shuttle Columbia disaster, NASA announced that the Shuttle was to be retired by 2010 and that the *AMS-02* was not on the manifest for any of the remaining Shuttle flights. Samuel had to lobby the United States

Congress and the public for permission to carry out one more Shuttle flight for the purpose of his project. Consequently, Samuel became the principal investigator for this international, $1.5 billion AMS experiment. Dr. Ting, the then 74-year-old Nobel laureate, said, "Real discovery is outside the ring of existing knowledge." For this project, Dr. Ting had to deal with endless technical problems in the fabrication and qualification of the large, sensitive, and delicate detector module. Eventually, the *AMS-02* was successfully launched on the Shuttle mission STS-134 on May 16, 2011, and installed on the International Space Station on May 19, 2011. This phenomenal undertaking involved over 500 scientists across 56 institutions and 16 countries. The experiment, if successful, could help NASA take a giant leap toward answering the question of what the universe is made of.

After five years of work on the International Space Station, AMS had already collected 80 billion cosmic rays, an amount much greater than that collected in the whole last century. A number of exciting, precise and unexpected results had been found from the study thus far. It is known that the collision of dark matter will produce energy which turns into ordinary matter, such as positron. The resulting excess of positrons can be accurately measured by the AMS residing in space. Such excess can be expressed as an increased positron fraction (i.e., the number of positrons divided by the summation of the numbers of positrons and electrons). The finding of abnormal positron fractions from the expected curve of collision of ordinary cosmic rays, was consistent with the presence of dark matter as the driving force. However, it would probably take a few more years to analyze further the already collected data in order to confirm the presence or absence of dark matter. Moreover, AMS also provided new data about the flux transitions of protons, helium, lithium, and other nuclei, which had contradicted the conventional understanding of cosmic ray behavior. These yet-to-be-analyzed data would help complete the study of dark matter and antimatter, as well as the search for new phenomena that would be beyond a scientist's imagination. As Dr. Ting emphasized, the role of basic science research might be viewed at the beginning by many to be very remote from human's daily activities. However, the importance of fundamental research should not be overlooked since such research undertakings can often provide the necessary foundations for the development of related, useful applications for the good of our society. Examples of such applications include television, laser, navigation, medical instruments, etc.[14]

Why the Discovery of the New J Particle Led to the Winning of the Nobel Prize

In fact, many new particles were discovered in the 1950s and 1960s. J. Robert Oppenheimer once marveled that a Nobel Prize could be given to any physicist who discovered a particle in any year. Protons and neutrons are not elementary; instead, they are made up of quarks, which are believed to be the elementary particles. Prior to Samuel's discovery, only three types of quarks were known: the up quarks, the down quarks, and the strange quarks. The importance of J/Psi centers on the fact that it is made up of a new type of quark called the charm quark. Such a discovery is considered much more important than finding other composite particles. Currently, six types of quarks have been discovered, namely, up, down, strange, charm, top, and bottom ones. These six types of quarks together with the electron, the muon, the tau, the electron neutrino, the muon neutrino, and the tau neutrino are the 12 elementary particles that form all matter that we know today. Elementary particles are so small that their individual structures still have not been discovered yet. Everyday life matter is made up of only three of the elementary particles: the up, the down and the electron. The others can only be created in high energy particle accelerators.

Personal Life

Dr. Ting married Susan Carol Marks in 1985 and they have one son, Christopher, who was born in 1986. Dr. Ting also has two daughters, Jeanne and Amy, from an earlier marriage.[5]

Other Accomplishments Achieved Over the Years

Dr. Ting's other important accomplishments include: the observation of nuclear antimatter (the anti-deuteron), the measurement of the size of the electron family (the electron, muon, and tau) and hence showing that the electron family has zero size (with a radius smaller than 10^{-17} cm); the precision study of light rays and massive light rays and thus illustrating that light rays and massive light rays can transform into each other at high energies, providing a critical verification of the quark model; performing precision measurements of the radius of the atomic nuclei; discovering the gluon (the particle responsible for transmitting the nuclear force); completing a systematic study of the properties of gluons; carrying out a precision measurement of muon charge asymmetry, being the first to demonstrate the validity of the Standard Electroweak Model (Weinberg, Glashow and Salam); determining the number of electron families and neutrino species in the universe; and precisely verifying the Electroweak Unification Theory. Lastly, Dr. Ting developed the first large superconducting magnet in space and has demonstrated the separation of helium isotopes in space.

Finally, below is the famous J-Particle sign.

J-Particle sign

(MIT General Collections. J-particle sign from Cyclotron building entrance [Building 44]). (Courtesy of MIT Museum.)

By the end of WWII, it was clear that in the future physicists would need access to increasingly larger and more expensive research facilities than any one university could provide. MIT and eight other major eastern universities formed a non-profit corporation to establish a new nuclear science facility. In 1947, under the direction of the U.S. Atomic Energy Commission, Brookhaven National Laboratory opened. The pioneering center has been home to seven Nobel Prize-winning discoveries. In 1974, during high energy particle physics experiments using Brookhaven's Alternating Gradient Synchrotron, MIT Professor Samuel C. C. Ting discovered a new kind of heavy elementary particle. Since the last meson was named the K particle, Ting named the new one the J particle. Ting won the 1976 Nobel Prize for this discovery, and gave his acceptance speech in Chinese, the first Nobel laureate to do so. The particle was independently and simultaneously discovered by a team at the Stanford Linear Accelerator Center whose leader, MIT alumnus Burton Richter, shared the prize with Ting. There were "J"s everywhere after the announcement of Ting's Nobel Prize, including this sign, which has remained above the entrance to the MIT Cyclotron on Vassar Street for 35 years. The sign is on loan from the MIT Department of Physics. [MIT 150 Exhibition label text; the above paragraph was obtained through the courtesy of the MIT Museum.]

Other Distinguished Awards

Ernest Orlando Lawrence Award (1975)
Eringen Medel (1977)
De Gasperi Award (1988)

Honors

Membership of US National Academy of Science, Academia Sinica, Russian Academy of Science and Pakistan Academy of Science.

Honorary Doctoral Degrees from:

Columbia University, The Chinese University of Hong Kong, Moscow State University, University of Science and Technology in China and the University of Bologna.

Editors' Note

During his Nobel Banquet Speech given on December 10, 1976, Dr. Ting, the first person who gave a speech in Chinese at a Nobel gathering, emphasized the following from his heart:

事实上，自然科学理论不能离开实验的基础、特别，物理学是从实验产生的。

我希望由于我这次得奖，能够唤起在发展国家的学生们的兴趣，而注意实验工作的重要性。

In reality, a theory in natural science cannot be without experimental foundations; physics, in particular, comes from experimental work.

I hope that awarding the Nobel Prize to me will awaken the interest of students from the developing nations so that they will realize the importance of experimental work.[4]

References and Recommended Readings

1. "Samuel C. C. Ting — Facts". *Nobelprize.org*. Nobel Media AB 2013. <http://www.nobelprize.org/nobel_prizes/physics/laureates/1976/ting-facts.html>.
2. "Samuel C. C. Ting — Biographical". *Nobelprize.org*. Nobel Media AB 2013. <http://www.nobelprize.org/nobel_prizes/physics/laureates/1976/ting-bio.html>.

3. "Samuel C. C. Ting — Nobel Lecture: The Discovery of the J Particle: A Personal Recollection". *Nobelprize.org.* Nobel Media AB 2013. http://www.nobelprize.org/nobel_prizes/physics/laureates/1976/ting-lecture.html.
4. "Samuel C. C. Ting — Banquet Speech". *Nobelprize.org.* Nobel Media AB 2013. <http://www.nobelprize.org/nobel_prizes/physics/laureates/1976/ting-speech.html>.
5. Samuel C. C. Ting. Wikipedia. Accessed on July 20, 2017.
6. Public Affairs Television. "Bill Moyers Special, 'Becoming American — The Chinese Experience'. Interview with Samuel Ting".
7. Samuel C. C. Ting. In: Franklin Ng, editor. The Asian American Encyclopedia. Marshall Cavendish, New York, N.Y., 1995, p. 1490.
8. Experimental Observation of a Heavy Particle J. Physical Review Letters 33 (23): 1404–6, 1974.
9. Alpha Magnetic Spectrometer-02 (AMS-02). NASA. August 21, 2009.
10. J. Hsu. Space Station Experiment to Hunt Antimatter Galaxies. Space.com. September 2, 2009.
11. A Costly Quest for the Dark Heart of the Cosmos. New York Times, November 16, 2010.
12. Samuel C. C. Ting, S. J. Brodsky. Timelike Momenta in Quantum Electrodynamics, Columbia University, United States Department of Energy (through predecessor agency the Atomic Energy Commission), December 1965.
13. Samuel C. C. Ting. Hadron and Photon Production of J Particles and the Origin of J Particles. Massachusetts Institute of Technology, International Conference on High Energy Particle Physics, Palermo, Sicily, Italy, June 23, 1975.
14. Samuel C. C. Ting, the J/psi Particle (Charm), and the Alpha Magnetic Spectrometer (AMS). Research and Development of the US Department of Energy 15th Anniversary Research and Development Accomplishment. webpage. http://www.osti.gov/accomplishments/ting.html.
15. Samuel C. C. Ting honored during "The Chinese University of Hong Kong 33rd Congregation (1987)". http://www.cpr.cuhk.edu.hk/cong/hongrads/343.
16. Samuel C. C. Ting. The Alpha Magnetic Spectrometer on the International Space Station. 2016 Series of Lectures on Astrophysics and Cosmology: science of the cosmos, science in the cosmos.

Photo 3.1. Samuel Ting poses with lens used in his 1961 PhD thesis experiment. Taken at the Ann Arbor Hands on Museum. (Photo credit: AIP Emilio Segrè Visual Archives, Lawrence W. Jones Collection.)

Photo 3.2. Samuel Ting and his research team at the Brookhaven National Laboratory. (Courtesy of the Brookhaven National Laboratory, USA.)

Photo 3.3. Samuel Ting receiving the Nobel Prize in Physics from King Carl XVI Gustaf of Sweden in Stockholm in 1976. (Photo credit: SCANPIX SWEDEN/Sipa USA.)

Photo 3.4. Samuel Ting accepting his 1976 Nobel Prize in Physics. (Photo by Jan Collsioo/Pressens Bild AB, courtesy AIP Emilio Segrè Visual Archives, Gift of Eleanor Dahl.)

Photo 3.5. Samuel Ting with fellow 1976 Nobel Prize winners. L-R are: D. Carleton Gajdusek, Milton Friedman, William Lipscomb, Samuel Ting, Burton Richter, Baruch S. Blumberg and Saul Bellow. (Photo by Leif R. Jansson/Svenskt Pressfoto, courtesy AIP Emilio Segrè Visual Archives, Gift of Eleanor Dahl.)

Photo 3.6. Samuel Ting with his daughters, Jeanne and Amy, at the 1976 Nobel Prize in Physics ceremony. (Photo by Jan Collsioo/Pressens Bild AB, courtesy AIP Emilio Segrè Visual Archives, Gift of Eleanor Dahl.)

Photo 3.7. Samuel Ting with fellow Nobel Laureates in Physics at the Brookhaven National Laboratory. Back row (left to right):1980 Laureates Val L. Fitch and James W. Cronin, as well as 1976 Laureate Ting. Front row (left to right):1957 Laureate Chen Ning Yang and 1944 Laureate Isidor Isaac Rabi. (Courtesy of Brookhaven National Laboratory.)

Photo 3.8. Samuel Ting with fellow Nobel Laureates in Physics at the Brookhaven National Laboratory in August 1996. (Left to right): 1957 Laureate Tsung-Dao Lee, 1988 Laureates Leon Lederman, Melvin Schwartz and Jack Steinberger, as well as 1976 Laureate Ting. (Courtesy of the Brookhaven National Laboratory.)

Photo 3.9. Samuel Ting receiving a Doctor of Science degree, *honoris causa*, from The Chinese University of Hong Kong at its 33rd Congregation in 1987. Photo taken with (from left) Prof. Hsu Baysung, Dr. the Hon. P. C. Woo, Prof. Gerald H. Choa, Dr. W. C. L. Brown, Laureate Samuel C.C. Ting, Sir David Akers-Jones, Dr. Leung Kau Kui, Dr. Lü Shu-xiang, Dr. the Hon. Q. W. Lee and Dr. Ma Lin. (Courtesy of The Chinese University of Hong Kong.)

Photo 3.10. March 30, 2011. CAPE CANAVERAL, Fla. — STS-134 Commander Mark Kelly, lying down, and Professor Sam Ting, Alpha Magnetic Spectrometer-2 (AMS) principal investigator at the Massachusetts Institute of Technology, perform a walkdown of space shuttle Endeavour's payload bay. The Express Logistics Carrier-3 packed with spare parts and the (AMS) are inside the bay for the STS-134 mission to the International Space Station. (Photo credit: Public domain/Wikimedia Commons.)

Photo 3.11. Samuel Ting (Front row, sixth from left) — who shared the 1976 Nobel Prize for Physics for discovering a new particle, now called "J/psi," confirming the existence of the charmed quark using Brookhaven Lab's Alternating Gradient Synchrotron — returned to the Lab on May 7, 2013 to present results from a particle physics detector aboard the International Space Station. (Courtesy of the Brookhaven National Laboratory.)

1943年，丁肇中和父母、弟弟、妹妹于重庆大学合影

Photo 3.12. Samuel with his parents, his younger brother and his sister at Chongqing University in 1943. (Photo courtesy of Wikimedia Commons; source of photo being History of Chongqing University library, 重慶大學校史馆.)

The badge of National Hsinchu Senior High School, the school being the alma mater of Laureate Lee.
(Source: Wikimedia Commons: 國立新竹高級中學 第61屆校友 楊翰傑 [Mr. HK Young, a Hsinchu high school alumnus]).

Motto of School: Honesty, Wisdom, the Glorious Health and Perseverance.

Laureate Lee donated his Nobel Prize medal to his alma mater.

Chapter 4

Yuan Tseh Lee, 李遠哲
1986 Nobel Laureate in Chemistry

Yi Zhou, Jian Sun and Guanhua H. Chen

Laureate Yuan Tseh Lee, advisor to Taiwan's presidential office, arrives for the Taiwan Energy Conference in Taipei, Taiwan, on April 15, 2009. Taiwan's President Ma Ying-jeou urged legislators to approve a law that encourages investment in renewable energy such as solar and wind power to help reduce greenhouse gas emissions blamed for climate change. (Courtesy of Maurice Tsai/Bloomberg via Getty Images.)

From Nobelprize.org:

Prize motivation: *"for their contributions concerning the dynamics of chemical elementary processes".*[1]

"I'll feel satisfied if my work can be recognized. As for the Nobel Prize, I've never thought it can have so deep influence on one's life,"[2] said Yuan Tseh Lee, who received the Nobel Prize in Chemistry in 1986 along with Dudley R. Herschbach and John C. Polanyi for their contributions pertaining to the dynamics of chemical elementary processes.[1,3]

He is hardworking: "If one thing has one hundred steps and if I do 5 percent better than others in every step, one hundred to the power of 105 percent means much better than others in the final." He is modest: "I just live a serious life." His co-worker Herschbach described him as "the Mozart of Physical Chemistry".[2] Maybe in life he did not show his talent as early as Mozart, but Yuan Tseh Lee was obviously a pioneer in the field of chemical dynamics.

Yuan Tseh Lee was born in Hsinchu City (新竹市), Taiwan on November 19, 1936. His father, Tze Fan Lee (李澤藩), was a famous painter and art teacher. His mother, Pei Tsai (蔡配), was the principal of a kindergarten school. Lee's parents were very strict with him. He was required to be careful and aspirant, which had subtle influence on the formation of Lee's character. During the Second World War, the Japanese occupied Taiwan. In the first two years of elementary school, as the city was ruined in the war, Lee followed the populace to take refuge in the mountains. There, he had a relaxing time playing in the countryside, and exploring nature.[4] In terms of sports, he was chosen as the second baseman for his school's baseball team. He also played a good game of Ping Pong. Even under heavy study pressure during high school, he still kept on playing tennis.[5] Lee characterized himself as an "Eager Beaver" (拼命三郎), "I don't care much about win or loss. I only try my best to play well".[2] Some years later, when he began his scientific research career, he kept this winning spirit and always finished his tasks in a satisfactory manner in a short time. Moreover, this gave him a healthy body to support his hardworking habit in his later years.

Lee was forced to speak Japanese during the Japanese occupation.[6] From the third grade of elementary school, he began to learn Chinese. Since then, he fell in love with reading. A variety of books provided him with knowledge in different areas. An article named 藍色的毛毯, (loosely translated by the present editors as "The Blue Woolen Blanket"), published on the 開明少年 (a magazine for enlightened young people), left a deep impression on him.[4] That article described the post-revolution transformation of Soviet Union's society into one of socialism. This experience made him realize that a society could be changed even though Taiwan was at the bottom of circumstances after the Second World War.

After graduating from elementary school, Lee went to Hsinchu High School (新竹高中). It was quite a fantastic period for him. The president of the school, Xin Zhi Ping (辛志平), was a great educator. Following the school's motto, Lee kept on improving himself, not only in knowledge but also in physical fitness and communication skills.[2] During high school, he came across the biography of Madam Curie. He got so motivated by her story that he decided to be a scientist like her: selfless, idealist and contributive.[6] Another item that is worth mentioning was that Lee was asked to assist his classmates in the study of natural sciences.[4] The preparation of review materials instilled into Lee the idea that one could learn faster and better by oneself. The fact that he embraced this idea had a deep influence on Lee's future career.

In 1955, Lee was admitted to the National Taiwan University (NTU) without the need to take the entrance examination due to his excellent performance in high school.[7] He chose to major in chemical engineering at first. At the end of his freshman year, however, he decided to switch to the field of chemistry to implement his interests in chemical research.[8] During the four years of university life, apart from taking chemistry courses, he also attended related courses in modern physics, including those in quantum mechanics, electrodynamics, etc. Those courses were thought by Lee to be important for his future pursuits in chemistry. He even learned Russian and thus was able to read papers from the Soviet Union directly.[9] During the summer of his freshman year, he and two other students remained in the dormitory and studied thermodynamics by themselves. Most of the time, he arrived at the university library by 7.00 a.m., waiting to enter, and stayed until closing time. Under such a free and exciting atmosphere in the National Taiwan University, Lee could focus on studying and research. In his final year of university life, he worked on the separation of strontium (Sr) and barium (Ba) using an electrophoresis method under the supervision of Professor Hua Sheng Cheng (鄭華生) and, as a result, was awarded his BS degree in Chemistry.[7]

After graduation, Lee continued his studies at the Nuclear Science and Technology Development Centre of National Tsinghua University and utilized his solid foundation in physics to further his postgraduate work. He received his Master's degree by studying the natural radioisotopes contained in Hukutolite (a type of barium sulfate, containing lead and traces of strontium, radium and calcium) under the guidance of Professor H. Hamaguchi.[8] After that, Lee stayed at the National Tsinghua University as a research assistant, helping Professor C. H. Wong to study the structure of tricyclopentadienyl samarium using X-ray techniques.

In the summer of 1962, Lee gained admission to the University of California at Berkeley (UCB) to pursue his doctoral degree under the supervision of Professor Bruce Mahan.[7] Even though Lee was equipped with a large measure of academic experience in his past, he found it difficult to adapt himself to this new environment at the beginning. However, he managed to overcome the sundry obstacles soon. The ability of self-learning again provided him with a lot of help. In addition, the equal relationship between student and supervisor offered him a lively atmosphere in terms of academic discussion. Given his ability to read Russian, he was able to solve a difficult scientific problem by consulting some articles published in Russian by Soviet scientists. Professor Mahan was quite satisfied with Lee's work and encouraged him to start preparing for his doctoral thesis earlier than usual.[4] After having received his PhD degree in 1965, Lee stayed at UCB and began to study the phenomenon of ion-atom collision. His interests in atom and ion matters could be traced back to his time at NTU. He always wanted to figure out how atoms and ions act during chemical reactions. From discussions with technical staff, he learned the art of designing and constructing very powerful apparatuses to carry out experiments such as the one involving: "$N_2^+ H^2 \rightarrow N_2H^+ + H$".[10] Using equipment designed by himself, Lee achieved a remarkable degree of success in discovering product distribution contour maps of a host of reactions.

In February 1967, Lee arrived at Harvard University as a post-doc fellow in Professor Dudley Herschbach's group. He studied the reactions between hydrogen atoms and diatomic alkali molecules with Professor Robert Gordon. At the same time, Lee tried to construct a universal crossed molecular beam apparatus with Professor Dong McDonald and Professor Pierre LeBreton. Due to time constraints, Lee pulled up his sleeves and participated in all the processes required in making the apparatus: from sketching the design to actual assembling. He seemed to have inherited his father's talent for painting. The sketches of the apparatus that he drew up were complex but clear-cut and beautiful. It was really surprising that the machine had been completed in only a period of 10 months, thus allowing the observation of the first non-alkali neutral beam experiment involving "$Cl + Br_2 \rightarrow BrCl + Br$" to be performed successfully in late 1967.

First demonstrated in 1953 by Taylor and Datz of Oak Ridge National Laboratory,[11] the crossed molecular beam approach was developed for the observation of chemical reactions. In this approach, two beams of atoms or molecules are collided together to elucidate the dynamics of the chemical reaction, while individual reactive collisions can be detected in the process. With their refined apparatus, Herschbach and Lee managed to probe into gas-phase reactions in unprecedented details.[12]

After the construction of his first apparatus, Lee went on to the University of Chicago in October 1968 as an assistant professor in the Department of Chemistry and the James Franck Institute. There, Lee remained creative and continued to improve his crossed molecular beam apparatuses. These state-of-the-art machines enabled him to carry out pioneering experiments and obtain exciting results with his students. His design even shocked the famous Professor R. B. Bernstein. When Bernstein visited the University of Chicago and saw what results Lee accomplished in an afternoon's work, he said that Lee had done what he had originally planned to do for the coming 10 years.[8] Due to his excellent work, Lee was promoted to the rank of associate professor and then to that of professor in 1971 and 1973, respectively. In 1974, he returned to UCB as a professor of chemistry and a principal investigator at the Lawrence Berkeley Laboratory of the University of California.

In January 1994, he returned to Taiwan to serve as the President of Academia Sinica. Apart from continuing his own chemical dynamics research, Lee made great efforts to promote student training as well as education improvement.[13]

Yuan Tseh Lee met his wife, Bernice Chin Li Wu (吳錦麗), in elementary school.[8] They have three children, Ted (son, born in 1963), Sidney (son, born in 1966) and Charlotte (daughter, born in 1969). Lee really appreciates the support from his wife and family. Without them, he believes that he could not have achieved so much.

Amidst a host of honors that Lee richly deserved, he once said, "Doing scientific research is to enjoy the academic life itself. Focusing on honor or profit can only limit one's achievement".[4] He never slowed down in spite of all the awards that came his way. Lee donated his Nobel Prize Medal to his alma mater, Hsinchu High School, as a means to express his gratitude to President Xin Zhi Ping of the school.

Finally, it is worth noting Lee's enlightening comment, "It's very satisfying to promote science and education and see good results. Setting a good example for young people, being a role model, is very important for me".[14] What a most heartwarming and exemplary statement emanating from a true academician, scientist and scholar!

Honors and Awards[15]

In addition to the Nobel Prize, his awards and distinctions include Sloan Fellow (1969); Fellow of the American Academy of Arts and Sciences (1975); Fellow Am. Phys. Soc. (1976); Guggenheim Fellow (1977); Member, National Academy of Sciences (1979); Member, International Academy of Science; Member, Academia Sinica (1980); E.O. Lawrence Award (1981); Miller Professor, Berkeley (1981); Fairchild Distinguished Scholar (1983); Harrison Howe Award (1983); Peter Debye Award (1986); National Medal of Science (1986). Yuan Tseh Lee was awarded the Othmer Gold Medal in 2008 in recognition of his outstanding contributions to progress in chemistry and science.

Editors' Note

The Editors cannot agree more with Laureate Lee's exemplary advice that, "Doing scientific research is to enjoy the academic life itself. Focusing on honor or profit can only limit one's achievement".[4]

References

1. Yuan T. Lee — Facts — Nobelprize.org https://www.nobelprize.org/nobel_prizes/chemistry/laureates/1986/lee-facts.html
2. Yang MY. (1995) Yuan Tseh Lee: The Mozart of Physical Chemistry. In: Global Views.
3. Yuan T Lee — Other Resources. *Nobelprize.org.* Nobel Media AB 2014. Web. 6 Dec 2016. <http://www.nobelprize.org/nobel_prizes/chemistry/laureates/1986/lee-or.html>
   Yuan T. Lee's page at Academia Sinica.
   'Yuan T. Lee and Molecular Beam Studies' from DOE R&D.
4. Lee YT. (1995) The Growing Up of a Scientist: Presentation at Tamkang University, Taiwan.
5. Su JX. (2003) Yuan Tseh Lee: Racket in hand power infinity. In: Min Sheng Daily Taiwan.
6. Lee YT. (1986) The Process of My Growth. He R.Y. ed.
7. Yuan T Lee — Biographical. In: Nobelprize.org: Nobel Media AB 2013.
8. Lee YTW, Cheng Wen, Li Yih Yuan. (1996) Seminar: Mentality of Academic Research, Sharing from Yuan Tseh Lee, Cheng Wen Wu, Yih Yuan Li. In: Central Daily News.
9. Common Wealth Magazine Editorial Department. (2004) Yuan Tseh Lee: The lonely boy beside the Touqian River. In: Common Wealth Magazine.
10. Gentry W, Gislason E, Mahan B, Tsao CW. (1967) Inelastic scattering of N_2^+ by helium. *J Chem Phys* **47**: 1856–7.
11. Taylor EH, Datz S. (1955) Study of chemical reaction mechanisms with molecular beams. The reaction of K with HBr. *J Chem Phys* **23**: 1711.
12. Lee YT. (1987) Molecular beam studies of elementary chemical processes. *Science* **236**: 793–8.
13. Lee YT. (2004) Some Reflections on Eduation Reform.
14. Guo YJ. (2004) Ten Years since Yuan Tseh Lee returned to Taiwan. In: Liberty Times.
15. Yuan Tseh Lee. Wikipedia, accessed on January 19, 2018.

Photo 4.1. Yuan Tseh Lee at Hsinchu High School, taken in the 1940s. (Courtesy of Laureate Lee.)

Photo 4.2. Laureate Lee's family portrait taken in 1954. Back row, second from left, Yuan Tseh; Front row from left to right, Lee's mother, Pei Cai and father ZF Lee. (Courtesy of Laureate Lee.)

Photo 4.3. Laureate Lee with his wife, Bernice Chin-li Wu, and his daughter Charlotte, taken in 1986. (Courtesy of Laureate Lee.)

Photo 4.4. Yuan Tseh Lee receiving the Nobel Prize from King Carl XVI Gustaf of Sweden in 1986. (Photo credit: Rolf Hamilton/PRESSENS BILD/TT/Sipa USA.)

Photo 4.5. Laureate Lee receiving a Doctor of Science, *honoris causa*, degree and delivering his presentation during the 38th Congregation of The Chinese University of Hong Kong in 1989. (Courtesy of The Chinese University of Hong Kong.)

Yuan Tseh Lee (李遠哲): 1986 Nobel Laureate in Chemistry

Photo 4.6. Laureate Lee receiving a Doctor of Science, *honoris causa*, degree from The University of Hong Kong in 2007. (Courtesy of The University of Hong Kong.)

Photo 4.7. Laureate Lee speaking at a press conference at the Queen Sirikit Convention Center in Bangkok on October 18, 2003. Lee, representing Taiwan, attended the APEC Leaders Summit on October 20–21, Bangkok, Thailand. (Courtesy of ADEK BERRY/AFP/Getty Images.)

Photo 4.8. Laureate Lee (Taiwan representative) sat next to Russian President Vladimir Putin during the "APEC Economic Leaders' Retreat 1" at Government House in Bangkok, on October 20, 2003. Leaders from the 21-member Asia Pacific Economic Cooperation (APEC) grouping met in the Thai capital on October 20–21. (Courtesy of SAM YEH/AFP/Getty Images.)

Photo 4.9. Laureate Lee, president of Academia Sinica, Taiwan's top state research institute, was escorted by Chilean President Ricardo Lagosfor to the Asia Pacific Economic Cooperation (APEC) Summit in Santiago, Chile, November 20, 2004. (Photo credit: Diego Giudice/Bloomberg/Getty Images.)

Photo 4.10. Dudley R Herschbach (left) and Yuan Tseh Lee (right) at John B Fenn (middle) 90 symposium, taken on June 15, 2007. (Photo credit: Allen Jones, VCU Creative Services.)

Photo 4.11. Laureate Lee was the President of Academia Sinica, Taipei, Taiwan from 1994 to 2006. (Courtesy of Academia Sinica.)

Photo 4.12. Laureate Lee playing tennis, in 2010. (Courtesy of Laureate Lee.)

Yuan Tseh Lee (李遠哲): 1986 Nobel Laureate in Chemistry

A girl is studying hard in the right lower corner of the photo. "The sea of knowledge is devoid of apparent shores; the real shores can only be reached through hard work (學海無涯, 唯勤是岸)." (Photo courtesy of Mr. Fan Ho.)

Chapter 5

Steven Chu, 朱棣文
1997 Nobel Laureate in Physics

S. M. Kurt Lee, Hon-Lok Tang, Ramin Sam
and Susie Q. Lew

Steven Chu, the 12th US Secretary of Energy, 2009–2013.

Nobel Prize motivation for Steven Chu (obtained from Nobelprize.org): "*For development of methods to cool and trap atoms with laser light*".[1]

The study of atoms in an isolated state offers valuable information about the fundamental properties of matter such as the quantum mechanics of the atomic components. However, atoms move at very high speeds. At room temperature they move at about, 500 m/sec. Steven Chu and two other research groups were able to devise an innovative approach to slow down these atoms to as little as 2 cm/sec and trap them in an isolated state for further examination and experimental measurements of their atomic structure.

In 1997, Steven Chu shared the Nobel Prize in Physics with Claude Cohen-Tannoudji and William D. Phillips. These three physicists were recognized for their work in developing methods to cool and trap atoms. Steven Chu is an accomplished researcher and his beautiful mind is characterized by talent, creativity, and logical thinking. These attributes were nurtured by his upbringing in an academically oriented environment.

Steven was born in St. Louis, Missouri in 1948 to an academic family. His father, Ju-Chin Chu (朱汝瑾), a graduate of Tsinghua University (清華大學), came to the US to study chemical engineering at the Massachusetts Institute of Technology (MIT) in 1943, and subsequently had a distinguished career as professor of chemical engineering. His mother, Ching-Chen Li (李靜貞), also a graduate of Tsinghua, studied economics at MIT too. His maternal grandfather, Shu-tian Li (李書田), earned a Ph.D. at Cornell University in 1926, and began his teaching career at Peiyang University (北洋大學堂). At the age of 28, he became president of Tangshan College of Engineering (唐山工程學院), and in 1932, was named President of Peiyang University. In 1951, Peiyang University was renamed Tianjin University (天津大學). Shu-tian's brother, Shu-hua Li (李書華) studied physics in France, and worked as a professor at Peking University (北京大學) and served as Minister of Education of China.

Steven was raised in Garden City, Long Island, New York and attended the Garden City public school while his father taught at the Brooklyn Polytechnic Institute. In his early grade school years, Garden City was a community of 25,000, including only two other Chinese families besides his own. Steven's family strongly emphasized education, and he and his brothers were told that the highest aspiration is to become a scholar and professor. His mother and father constantly encouraged the brothers to read. Steven's elder brother, Gilbert, earned the highest cumulative average in the Garden City High School's recorded history. He went on to receive a Ph.D. in physics from MIT and an M.D. from Harvard. While growing up, both Steven and his younger brother, Morgan, did not want to compete with their older brother in high school. Nevertheless, the two younger brothers began to excel in college and beyond. The younger brother, Morgan, earned a Ph.D. at the age of 22, followed by a law degree, and became a leading patent attorney.

Several outstanding features of Steven's early schooling sowed the seed that would influence his development as a brilliant research scientist. In grade school he "constructed devices of unknown purpose", built homemade rockets, and model airplanes. He started a neighborhood "business" with a high school friend offering to test their neighbor's soil to optimize lawn fertilizer mixtures. He was encouraged to work with his hands and "picture things geometrically". In a 2004 television interview, Steven suggested that these early experiences helped him develop the "spatial" intuition necessary to design the instruments that he later used to conduct his atomic experiments.

In addition to the informal, "hands-on" training, Steven emphasized his appreciation of a particular 9th grade geometry teacher. Instead of memorizing facts, he was introduced to the concept of logical thinking. He learned about the elegance of mathematics, "beginning from a few intuitive postulates, far reaching consequences could be derived". In his senior year of high school, Steven took advanced placement courses in physics and calculus. He had a gifted physics teacher who was nationally recognized. In reminiscing his influence, Steven recalled in his Nobel biography, "He told us we were going to learn how to deal with very simple questions such as how a body falls due to the acceleration of gravity. Through a combination of conjecture and observations, ideas could be cast into falsifiable theories that can be tested by experiments. These theories, held accountable to continued experimental challenges, would either have to be refined or abandoned. Despite the modest goals of physics, knowledge gained in this way would become collected wisdom that could not be discarded by changing fashion. Progress became cumulative through the ultimate arbitrator — experiment". While in his senior year of high school, Steve constructed a pendulum to make a precision measurement of gravity. Years later in his Nobel Prize winning research, he showed that the atom interferometer with laser-cooled atoms became the most precise method of measuring acceleration due to gravity.

Steven enrolled at the University of Rochester in New York as an undergraduate student, majoring in physics and mathematics. His two-year introductory physics sequence used **The Feynman Lectures in Physics** as the textbook. The lectures, and especially the ones that departed from the standard textbook approaches to physics cemented his love of physics. In fact, the Feynman lectures and problem sets are famous for fostering the skills of "lateral thinking", which is one of the main themes of creative and critical thinking.

In 1970, Steven entered the graduate program in physics at the University of California at Berkeley (UCB). UCB was a world center of physics that included seven active Nobel Laureates. He was interested in high-energy particle physics, but was not drawn to the large collaborations. Instead, he decided to work in the very small group of Professor Eugene Commins. The first project that Gene Commins assigned him was a theoretical astrophysics problem, but after a few months, Steven found that he was much more drawn to experimental physics. In a television interview in 2004, Steven recalled that Commins would work alongside his students like a colleague and had the ability to make all of his students feel special.

In 1978, after his time as a graduate student and postdoctoral fellow at Berkeley, Steven joined Bell Laboratories, another world-leading research institution. There, he was exposed to an incredibly rich but different research atmosphere. Bell Laboratories was an extremely idea-rich environment, and where there was much less stove piping of research into traditional scientific disciplines. When he first joined, he was encouraged to explore possible areas of research instead of settling quickly into more familiar areas.

It was at Bell Labs that Steven was introduced to Art Ashkin (Editors' Note: For his long-standing, exemplary research work on "the optical tweezers and their application to biologic systems", Dr Ashkin was awarded the Nobel Prize in Physics in 2018). Art and his collaborators did a number of seminal experiments demonstrating the optical forces that one can exert on atoms, but after more than eight years of work, the trapping of atoms with light did not appear imminent, and the research program was shut down. Rather than trying to trap the atoms directly, Steven's approach was to use a laser to cool the atoms to very low temperatures with three pairs of counter-propagating laser beams arranged at right angles to one another. Once the atoms were cooled from hundreds of degrees Kelvin to less than a thousandth of degree above absolute zero, it became relatively easy to trap them. The initial optical trap based on a single focused laser beam was followed by the demonstration of the magneto-optic trap that rapidly became the workhorse optical trap in the following three decades.

Another critical step was made by William Phillips who discovered that the laser cooling scheme first demonstrated at Bell Labs cooled atoms to temperatures much lower than what was believed to be the theoretical limit. To explain the new cooling mechanism, Claude Cohen-Tannoudji and Steven Chu proposed that there was an additional cooling effect for the atoms based on the changing electric field polarization in the optical molasses geometry; this effect became known as "optical pumping".

The first application of laser cooling was the creation of an atomic fountain clock that rapidly led to an improvement in the world time standard. Other applications in atomic physics, such as Steven's atom interferometers, rapidly followed. In 1991, he used the optical tweezers trap (first used to trap atoms) to hold onto individual molecules of DNA. This work helped open up the world of single molecule research in biology. In 1997, Steven shared the Nobel Prize in Physics with Claude Cohen-Tannoudji and William Daniel Phillips for developing methods to cool and trap atoms.

On February 13, 2004, in an interview at the University of California, Berkeley, Steven was asked about his research and what were the qualities to do well in physics research. 'Apart from a strong mathematics background, a successful scientist typically has a natural curiosity, a strong internal drive and a passion to find out the "answer" to the scientific question they posed. Also, a critical "doggedness" in the pursuit of the scientist's goal is highly correlated with scientific success.'

'A strong internal rudder is also useful when making a discovery that goes against the prevailing scientific view. While experimental science remains the ultimate arbitrator, one reaction to a new discovery is that the result is "wrong". As an understanding of the discovery deepens, the understanding becomes more intuitive, and the second reaction is that the new understanding was "trivial". Finally, there are always precursors to any discovery; the final reaction is that you are not the first person to make the discovery. However, one should not be discouraged and should continue to move on and find the truth. The path of scientific discovery is full of scientific disappointments, whereas the "Eureka" moments are rare.' Steven's ultimate success is due largely to his persistence, his desire to find the truth, and his remarkable vertical and lateral thinking skills developed in his youth; all these qualities were built on the foundation of strong family values.

Steven Chu received an honorary degree from Harvard University in 2009. During his speech, he said, "I began teaching with the idea of giving back; I received more than I gave". With the hope of inspiring the young generation in China to strive for excellence, he agreed to have the newly built primary school in Taicang (太倉市), Jiangsu Province (江蘇省), named after him. When he visited the school in 2000, he wrote, "Primary school is the first school, and as the first school, it is the first important step forward in a lifelong quest for knowledge. May the students learn to love learning during this first step".

In January 2009, Steven became the 12th Secretary of Energy of the United States of America. He shepherded admirably the energy program of the Obama administration until April 2013. With the belief that to combat climate change, the human emission of greenhouse gases needs to be drastically reduced, Secretary Chu advocated for more research into energy efficiency, renewable energy and nuclear power. He encouraged science students to take on this crucial environmental challenge, and participate in environmental planning and global initiatives. In his resignation announcement to Energy Department employees, he said, "As the saying goes, the Stone Age did not end because we ran out of stones; we transitioned to better solutions".

Some Representative Awards and Recognitions

Apart from being a co-winner of the 1997 Nobel Prize in Physics for his work on laser cooling for atoms, Steven Chu also received myriad other awards, was presented with 30 honorary degrees from universities around the world, and has been elected to the National Academy of Sciences, the American Philosophical Society, the American Academy of Arts and Sciences, the National Academy of Inventors, the Academia Sinica, and is a foreign member of the Royal Society, the Royal Academy of Engineering, the Chinese Academy of Sciences, and the Korean Academy of Sciences and Technology.

References

1. The Nobel Prize in Physics 1997 — Nobelprize.org www.nobelprize.org/nobel_prizes/physics/laureates/1997/
2. Steven Chu — Facts — Nobelprize.org http://www.nobelprize.org/nobel_prizes/physics/laureates/1997/chu-facts.html. Accessed on June 20, 2015.
3. Steven Chu — Biographical. From The Official Web Site of the Nobel Prize. Nobelprize.org. http://www.nobelprize.org/nobel_prizes/physics/laureates/1997/chu-bio.html. Accessed November 26, 2013.
4. Conversation with history, Institute of International Studies, UC, Berkeley, Webcast on February 13, 2004.
5. Dr. Steven Chu. Department of Energy www.energy.gov/contributors/dr-steven-chu
6. Steven Chu. Department of Physics https://physics.stanford.edu/node/2023
7. Steven Chu. From: Famous Scientists. The Art of Genius. www.famousscientists.org/steven-chu/
8. Steven Chu. Wikipedia. Accessed on April 6, 2017.
9. Steven's father, Ju-Chin Chu (朱汝瑾):
    (a) Ju-Chin Chu. http://www.digplanet.com/wiki/Ju-Chin_Chu (Wikipedia, accessed on September 17, 2016).
    (b) 互動百科. http://www.baike.com/wiki/朱汝瑾 (accessed on November 26, 2013).
10. Steven's mother, Ching-Chen Li (李靜貞): 美國能源部長朱棣文之母-李靜貞. 百度百科. http://baike.baidu.com/view/2599377.htm (accessed on November 26, 2013). Please note that Li is the surname.
11. Steven's maternal grandfather, Shu-tian Li (李書田):
    (a) Shu-tian Li. https://en.wikipedia.org/wiki/Shu-tian_Li (Wikipedia, accessed on September 17, 2016).
    (b) Shu-tian Li remembered. 回憶朱棣文的外祖父李書田校長. 歷史論壇. http://bbs.lishi5.com/thread-41183-1-1.html Accessed on November 26, 2013.
12. Steven's maternal grand uncle, Shu-hua Li (李書華): Shu-hua Li. https://en.wikipedia.org/wiki/Li_Shu-hua (Wikipedia, accessed on September 17, 2016).

Photo 5.1. Steven Chu, his son Michael and his mother Ching Chen Li (李靜貞). (Courtesy of Laureate Chu.)

Photo 5.2. Steven Chu's father, Dr. Ju-Chin Chu (朱汝瑾) was highly honored, in 1998, by the Chinese-American Engineers and Scientists Association of Southern California (CESASC). (Courtesy of Chinese-American Engineers and Scientists Association of Southern California.)

Photo 5.3. Steven Chu, left, with younger brother, Morgan, middle, and older brother, Gilbert, right. (Courtesy of Laureate Chu.)

Photo 5.4. Garden City High School. (Courtesy of Garden City Public Schools.)

Photo 5.5. Steven Chu at the Bell Laboratories in his early thirties. He began his research on the cooling and trapping of atoms with a laser light. He spent 9 years working there. (Courtesy of Laureate Chu.)

Photo 5.6. Steven Chu and his wife, Jean. (Courtesy of Laureate Chu.)

Photo 5.7. Steven Chu receiving the Nobel Prize from King Carl XVI Gustaf of Sweden in 1997. (Photo credit: Ulf Palm/TT/Sipa USA.)

Photo 5.8. Steven Chu and the US President, Barack Obama, in the US Department of Energy.

Photo 5.9. Steven Chu with the Purdue University's 2011 Solar Decathlon team in front of the solar-powered house that the team built for the competition.

Photo 5.10. The Steven Chu Primary School in Taicang, Jiangsu Province, China. (Courtesy of Ms. Liling Luk, principal of Steven Chu Primary School.)

Photo 5.11. Laureate Steven Chu and his wife, Jean. Taken with a student during their visit to the Steven Chu Primary School in 2000. (Courtesy of Ms. Liling Luk, principal of Steven Chu Primary School.)

Photo 5.12. Laureate Steven Chu on the front cover of a popular Chinese magazine. The title of that particular magazine issue was: "The Super Star across the Ocean". (Courtesy of Laureate Chu and the Beijing Jiaotong University Press.)

Steven Chu (朱棣文): 1997 Nobel Laureate in Physics

In 2010, Laureate Tsui paid a visit to Peking University's International Center of Quantum Materials (ICQM). Subsequently, the Daniel C. Tsui Laboratory was established as a component of the ICQM in 2012. (Courtesy of Peking University.)

Chapter 6

Daniel Chee Tsui, 崔琦
1998 Nobel Laureate in Physics

Huai-Bin Zhuang, Xin-Cheng Xie
and Fu-Chun Zhang

Daniel Chee Tsui (崔琦), 1998 Nobel Laureate in Physics. (Courtesy of the World Scientific Publishing Co.)

"Doing homework is usually like a sightseeing trip to resorts that have already been fully developed, whereas research is like an adventure to an intriguing destination in the trackless wilderness."[1] This perspective was offered to young students by Daniel Chee Tsui, who received the 1998 Nobel Prize in Physics alongside Horst L. Störmer and Robert B. Laughlin for their "discovery of a new form of quantum fluid with fractionally charged excitations" (Daniel C. Tsui — Facts — Nobelprize.org).

The journey of research described by Tsui is somehow similar to the journey of his early life while growing up. Both journeys appear to be full of excitement and uncertainty, where no one knows how far away the destination truly is and how difficult the travel can be. If one tries to understand how Tsui managed to achieve happiness and success against myriad hardships in both journeys, there are two quotes from him that might shed some light: "The only meaningful life is a life of learning,"[2] and "The most important part of life is the relationships with people."[1] The former quote explains the dedication and loyalty to science that Tsui has displayed during his search for knowledge while staying away from the spotlight. The latter embodies his openness, kindness, and humor with others, thus bringing about both a closely knit family and strong friendships with other people, including other scientists. Genuine, long-lasting, collaborative working partnerships with other scientists do carry some weight in modern scientific careers. In Tsui's growth into a true scientist and a respected person, virtues such as his clear visions and tireless efforts were the essential ingredients that enabled him to overcome obstacles and to leave a mark in the educational and scientific world.

Early Life in Fan Village, China

Tsui was born in February of 1939 in Fan Village (范莊村), which was a less developed area in Baofeng Township (寶豐縣), Henan Province (河南省) in central China. His father Changsheng Tsui (崔長生), like most other villagers at the time, had no opportunity to learn how to read and write. His mother, Shuangxian Wang (王雙賢), was not allowed to be literate because of the prevailing traditional gender inequality against women. Nevertheless, being born into an intellectual family well known locally, Wang acquired a broad vision and an amiable character.[3] Tsui's parents, especially his mother, were very determined to make sure that their children would receive the best possible education, no matter what the costs were. After years of war, unfortunately, the reconstruction of the education system in mainland China turned out to be extremely arduous and did not proceed as planned. Tsui finished primary school at age 10, but did not have access to middle school until age 12. His parents eventually made a tough decision to send Tsui to Hong Kong in pursuit of a better education. Sadly, Tsui never had a chance to see his parents again after having moved to Hong Kong, as his parents both passed away sometime in the 1950's and 1960's.

Life in Hong Kong and Pui Ching Middle School

Going to Hong Kong in 1951 was undoubtedly one of the most important steps in Tsui's life. After leaving his parents shortly after the lunar Chinese New Year,[4] Tsui first made a stop in Beijing to see his eldest sister Ying Tsui, then he headed off to Hong Kong to meet his two other sisters, Lou Tsui and Ko Tsui (also known as Au Tsui Wong). In Hong Kong Tsui began his adolescence with schooling at the sixth grade level, struggling with learning the Cantonese dialect and coping with loneliness, homesickness, limited financial resources, and long commutes each day to and from school. Nevertheless, some of his best virtues took shape during this time period, such as his perseverance, his modesty, his generosity, his love of helping others, and his strong sense of humor.[3,5] Hong Kong was the place where he became conscious of the joy of learning and where he started to form strong, lasting bonds with others. To Tsui, Hong Kong felt more like a home. He once said, "Home is a feeling of warmth rather than a place."[1]

In 1952, Tsui entered Pui Ching Middle School (培正中學) in Hong Kong. As a private Chinese school first founded in Guangzhou in 1889, Pui Ching Middle School (Hong Kong) has been celebrated for achievements in natural science subjects since the 1930's and for its successful graduates, including Daniel Tsui, Shing-Tung Yau, Lu Jeu Sham and Alfred Y. Cho.[6] The teachers were so outstanding, even overqualified according to Tsui, that had it not been for the upheaval of war, they would have been highly accomplished scholars and scientists. It was their intellect and vision that deeply inspired Tsui, who said "living in a commercialized city, to look beyond the dollar sign and see the exploration of new frontiers in human knowledge is an intellectually rewarding and challenging pursuit."[2] To pay tribute, Tsui donated his Nobel Prize Medal to Pui Ching in 1999.[7]

Tsui decided to enroll in the Special Classes Center, which prepared top-notch students from Chinese-medium schools for entry to The University of Hong Kong, although he had already been admitted to the medical school of National Taiwan University in Taiwan after his graduation with distinction from Pui Ching in 1957. In October, 1957, C. N. Yang and T. D. Lee were awarded the Nobel Prize in Physics. The news became a tremendous inspiration for a brighter future for Tsui and other young Chinese students, whose childhoods were affected by the dark shadow of wars. Tsui began to set his sights on entering the realm of physics to follow the giant steps of his role models, Yang and Lee. Tsui wanted to pursue research training at the University of Chicago, just like Yang and Lee did. Despite being in conditions that were far from ideal, Tsui was equipped with a burning passion to learn and to gain more knowledge of the world, a disposition paved by his time at Pui Ching. Tsui felt a satisfying sense of curiosity and exploration. Due to a lack of access to libraries, Tsui would stand straight for hours in Chinese bookstores just to read books. To cope with the shortage of teachers in the sciences at the Special Classes Center, Tsui and several classmates managed to learn physics and chemistry mainly by self-studying. His classmates reminisced that Tsui would often begin the day by saying, "Isn't it marvelous that …" and continue on to describe a fascinating phenomenon which he had just learned about the night before.[8]

Augustana College, Rock Island, Illinois

In the late spring of 1958, a surprising but wonderful news arrived: Tsui was admitted with a full scholarship to his church pastor's Lutheran alma mater, Augustana College, in Illinois. This signified Tsui's entry into America. Augustana College features a liberal arts curriculum and close interactions between faculty and students, given its small class sizes. Armed with a strong drive, hardworking habits, independent study skills, and experiences in Hong Kong, adjusting to a new environment did not appear to be a problem for Tsui during this transition. Tsui distinguished himself not only in the classroom, but also in the social life of the campus community. His academic attainments earned him the college's most coveted prizes, including election into Augustana's prestigious Phi Beta Kappa chapter. Other students responded to his strong character, proven by Tsui's winning of the Mr. Friendship award, given to an admired, trusted and highly valued friend to all, during his graduation in 1961. Tsui had actually graduated a year earlier than most students would.[9]

University of Chicago, Chicago, Illinois

As he had always dreamt about, Tsui became a graduate student at the University of Chicago. He realized very early on that he wanted to do tabletop experiments. Working closely with his thesis advisor, Royal Stark, who was a young, energetic professor, Tsui received the opportunity to learn from the bottom up — from basic machining processes to construction of laboratory apparatuses. Tsui also strengthened his enthusiasm for doing basic research collaboratively by matching Stark's intense pace of working: often 16–18 hours a day, for seven days a week. Stark was most impressed by Tsui's instincts for always asking intelligent questions and for looking into the experimental results that contradict expectations, which graduate students usually tend to ignore. By the time Tsui finished his thesis research, Stark had already considered Tsui to be a colleague rather than a student.[10] Moreover, what made Chicago more special to Tsui was that he met his wife, Linda Varland there. They married in 1964. They had two daughters, Aileen and Judith, and together, the family was an oasis of hope and support for Tsui.[11]

Bell Labs, Murray Hill, New Jersey

Tsui left Chicago in the early spring of 1968 to take up a position at the Bell Laboratories in New Jersey to do research on solid state physics. He investigated two-dimensional electrons, which was a new frontier in research, outside mainstream semiconductor research. In the late 1970's when Horst L. Störmer joined the Bell Laboratories, Tsui was already recognized as one of the world's leading experts on two-dimensional electron systems in silicon and was willing to share ideas, data and credit freely with others. Since Störmer's joining the Lab, the decades-long collaboration and friendship between the two blossomed.[11,12] In 1981, Tsui and Störmer applied extremely low temperatures and powerful magnetic fields to the high-quality, two-dimensional interface of the sandwiched semiconductor gallium-arsenide/aluminum-gallium-arsenide (GaAs/AlGaAs) wafers. In the extreme conditions where strong electron correlation is expected, they initially planned to look for signs of the theoretically predicted "electron crystal," but instead surprisingly discovered the fractional quantum Hall effect (FQHE), leaving a new phenomenon for the world to savor and explain. Because of the superior caliber of this work, Tsui won the Nobel Prize at the age of 43, along with Störmer.[13]

First explained by Robert Laughlin in 1983 and now understood, FQHE[14,15] is a fundamentally new macroscopic manifestation of the quantum laws, where electrons in extreme conditions behave collectively with new particles emerging at one third of the electric charge of an electron. The finding opened a new stream of research and today still continues to stimulate new discoveries in many fascinating fields, such as approaches to error-free topological quantum computation.

Princeton University, Princeton, New Jersey

In February 1982, soon after the FQHE discovery, Tsui moved to Princeton University and started teaching. He supervised postdocs and students to work on diverse topics on the two-dimensional electron system, and is highly applauded for his ability to make many of those around him to consistently produce high quality science and to eventually establish themselves as highly regarded scientists.[16]

Before Tsui received the Nobel Prize, he was honored by many other awards, such as the Oliver Buckley Prize (1984) and the Benjamin Franklin Medal (1998). Nevertheless, on the morning of the Nobel Prize announcement when the press corps fired away with one question after another, Tsui smiled and said, "You must not take this too seriously — life goes on."[17] Indeed, true to his characteristic modesty, what else could matter more than always moving forward in a life brimming with dedicated learning and whole-hearted sharing?

Honors[18]

Tsui is a member of the United States National Academy of Sciences, a member of the National Academy of Engineering (2004 election), a fellow of the American Association for the Advancement of Sciences, and a fellow of the American Physical Society. In 1992, Tsui was elected Academician of Academia Sinica, Taipei. In June 2000, Tsui was elected Foreign Member of the Chinese Academy of Sciences, Beijing.

Editors' Note:

1. The following sentences lifted from the above text richly deserve our utmost heart-wrenching emphasis:

   "It was their (the editors' italics: *the Pui Ching teachers*') intellects and visions that deeply inspired Tsui who said 'living in a most commercialized city (the editors' italics: *Hong Kong*), to look beyond the dollar sign and see the exploration of new frontiers in human knowledge as an intellectually rewarding and challenging pursuit."[2] To pay tribute, Tsui donated his Nobel Prize Medal to his alma mater, Pui Ching Middle School, in 1999.[7]

   Editors' Note: Laureate Tsui had it right. Money is not everything in life!

2. The Daniel C. Tsui Laboratory at Peking University, named after Professor Daniel C. Tsui, was established in June 2012. Professor Daniel C. Tsui is the honorary director, and Professor Ruirui Du serves as the executive director. This laboratory aims to research the quantum transport properties and spectroscopy of electrons at ultra-low temperatures, in low-dimension quantum materials, and using mesoscopic devices.

3. On November 23, 2017, Laureate Tsui was greatly honored with the inauguration of an educational facility, the Daniel Tsui Hall (崔琦教學大樓), by the Guangxi Peixian International College (培賢國際職業學院) located at Pingguo County, Baise, *Guangxi* Zhuang Autonomous Region, China (中國, 廣西壯族自治區, 百色, 平果縣). This exemplary facility should serve to inspire generations of students to strive for excellence, a noble educational goal so fervently championed by Laureate Tsui.

References

1. *Outstanding Chinese — Daniel C. Tsui*, Radio Television Hong Kong, telecasting on December 24, 2005.
2. Tsui DC. (2007) Daniel C. Tsui — Autobiography. In: Benette R and Joshi PC (eds), *The Encyclopaedia of Nobel Laureates: Physics*, 2007, p. 138. Dominant Publishers & Distributors, Privated Limited, New Delhi.

3. Tsui Wong A. (1999) My Brother, Dan Tsui. In: Wong C-Y, Lo JS-I and Lo SY (eds), *The Joy of the Search for Knowledge: A Tribute to Professor Dan Tsui*, p. 21. World Scientific Publishing Co. Pte. Ltd., Singapore.
4. Tsui DC. (2012) Speech given during the ceremony related to the conferring of a Peking University Honorary Doctoral Degree to Professor Daniel C. Tsui, on May 29, 2012.
5. Lo JSI. (1999) The Great Joy of Learning. In: Wong C-Y, Lo JS-I and Lo S-Y (eds), *The Joy of the Search for Knowledge: A Tribute to Professor Dan Tsui*, p. 49. World Scientific Publishing Co. Pte. Ltd., Singapore; Fein B. (1999) Pui Ching Reminiscences, *ibid*, p. 58.
6. Lam YH. (1999) Former Pui Ching Principal on Dan Tsui Winning the Nobel Prize. In: Wong C-Y, Lo JS-I and Lo S-Y (eds), *The Joy of the Search for Knowledge: A Tribute to Professor Dan Tsui*, p. 32. World Scientific Publishing Co. Pte. Ltd., Singapore; Hu B. (1999) On Education at Pui Ching. *ibid*, p. 70.
7. http://www.puiching.edu.hk/schoolevent/1999/DanielTsui/, *Daniel Chee Tsui revisits Pui Ching*, Pui Ching Middle School at Hong Kong, viewed on August 30, 2013.
8. Wong CY, Cheung FFK. (1999) Dan Tsui at the Special Classes Center in Hong Kong in 1957/1958, In: Wong C-Y, Lo JS-I and Lo S-Y (eds), *The Joy of the Search for Knowledge: A Tribute to Professor Dan Tsui*, p. 74. World Scientific Publishing Co. Pte. Ltd., Singapore.
9. McLaughlin DE. (1999) Dan Tsui at Augustana College, 1958–1961: Recollections of a Math Teacher. In: Wong C-Y, Lo JS-I and Lo S-Y (eds), *The Joy of the Search for Knowledge: A Tribute to Professor Dan Tsui*, p. 87. World Scientific Publishing Co. Pte. Ltd., Singapore; Benson TL. (1999) Reflection on Dan Tsui's Years at Augustana College. *ibid*, p. 99.
10. Stark RW. (1999) An Open Letter to Daniel C. Tsui. In: Wong C-Y, Lo JS-I and Lo S-Y (eds), *The Joy of the Search for Knowledge: A Tribute to Professor Dan Tsui*, p. 103. World Scientific Publishing Co. Pte. Ltd., Singapore.
11. Tsui L. (1999) My husband, Dan Tsui. In: Wong C-Y, Lo JS-I and Lo S-Y (eds), *The Joy of the Search for Knowledge: A Tribute to Professor Dan Tsui*, p. 24. World Scientific Publishing Co. Pte. Ltd., Singapore.
12. Störmer HL. (2007) Horst L. Störmer — Autobiography. In: Benette R, Joshi PC (eds), *The Encyclopaedia of Nobel Laureates: Physics*, p. 132. Dominant Publishers and Distributors, New Delhi.
13. Tsui DC, Störmer HL, Gossard AC. (1982) Two-Dimensional Magnetotransport in the Extreme Quantum Limit. *Phys Rev Lett* **48**: 1559.
14. Prange RE, Girvin SM. (eds) (1990) The *Quantum Hall Effect, 2nd ed.* Springer New York.

15. Chakraborty T, Pietiläinen P. (1988) *The Fractional Quantum Hall Effect*. Springer Berlin Heidelberg.
16. Chang AM. (1999) Prof. Dan Tsui — The Consummate Scientist. In: Wong C-Y, Lo JS-I and Lo S-Y (eds), *The Joy of the Search for Knowledge: A Tribute to Professor Dan Tsui*, p. 143. World Scientific Publishing Co. Pte. Ltd., Singapore.
17. Wei J. (1999) My Colleague Dan Tsui. In: Wong C-Y, Lo JS-I and Lo S-Y (eds), *The Joy of the Search for Knowledge: A Tribute to Professor Dan Tsui*, p. 127. World Scientific Publishing Co. Pte. Ltd., Singapore.
18. Wikipedia: Daniel C. Tsui, accessed November 14, 2016.
19. Daniel C. Tsui — Biographical — Nobelprize.org www.nobelprize.org/nobel_prizes/physics/laureates/1998/tsui-bio.html
20. Daniel C. Tsui — Facts — Nobelprize.org https://www.nobelprize.org/nobel_prizes/physics/laureates/1998/tsui-facts.html
21. Daniel C. Tsui — Nobel Lecture: Interplay of Disorder … — Nobelprize.org www.nobelprize.org/nobel_prizes/physics/laureates/1998/tsui-lecture.html
22. Daniel C. Tsui receives his Nobel Prize — Media Player at Nobelprize.org www.nobelprize.org/mediaplayer/index.php?id=931
23. Watch a video clip of the 1998 Nobel Laureate in Physics, Daniel C. Tsui, receiving his Nobel Prize medal and diploma during the Nobel Prize Award Ceremony. The Nobel Prize in Physics 1998 — Nobelprize.org www.nobelprize.org/nobel_prizes/physics/laureates/1998/
24. Ho H-W. (1999) Dan Tsui at the New York Celebration. In: Wong C-Y, Lo JS-I and Lo S-Y (eds.), *The Joy of the Search for Knowledge: A Tribute to Professor Dan Tusi*, p. 160. World Scientific Publishing Co. Pte. Ltd., Singapore.

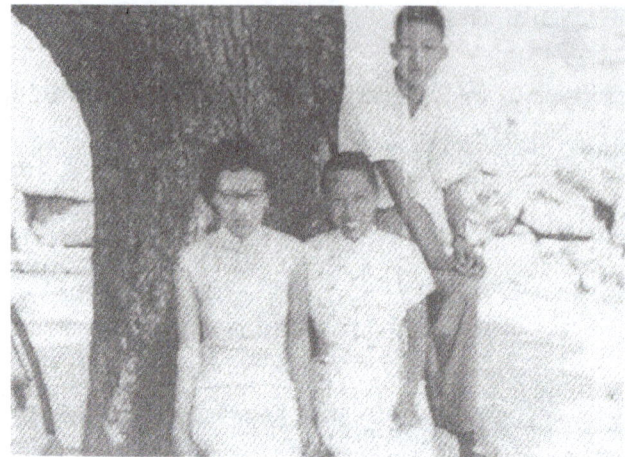

Photo 6.1. Dan Tsui and his two sisters in Hong Kong in 1950's. From left to right: Lou Tsui, Au Tsui Wong and Dan Tsui. (Courtesy of the World Scientific Publishing Co.)

Photo 6.2. Pui Ching Middle School in Hong Kong where Laureate Tsui received his secondary education. (Courtesy of Dr. Hon-Lok Tang.)

Photo 6.3. Dan Tsui (second left in the back row) with classmates in a Pui Ching Middle School alumni reunion in 1981. (Courtesy of the World Scientific Publishing Co.)

Photo 6.4. Dan Tsui receiving his Nobel Prize in Physics from the King of Sweden, Carl XVI Gustaf, in 1998. (Photo credit: Jonas Ekströmer/Scanpix Sweden/Sipa USA.)

Photo 6.5. Nobel Laureates in Physics in 1998: Horst L. Störmer (left), Robert B. Laughlin (middle) and Daniel C. Tsui (right) with their Nobel Diplomas and Medals. (Photo credit: TT News Agency/AFLO.)

Photo 6.6. Laureate Tsui with fellow 1998 Nobel laureates in Stockholm. Standing row, from left: Horst L. Störmer and Robert B. Laughlin, (Physics); Amartya Sen, (Economics); John A. Pople, (Chemistry); Ferid Murad, (Medicine). Sitting row: Louis J. Ignarro, (Medicine); Daniel C. Tsui, (Physics); Jose Saramago, (Literature); Robert F. Furchgott, (Medicine). (Photo credit: JONAS EKSTROMER/AFP/Getty Images.)

Photo 6.7. Laureate Tsui receiving the degree of Doctor of Science, *honoris causa*, from The University of Hong Kong in 2007. (Courtesy of the University of Hong Kong.)

Photo 6.8. Laureate Tsui (third from the right) with Pro-Chancellor the Hon. David Li Kwok Po (fourth from the right) and Vice-Chancellor Lap-Chee Tsui (first from the right) of The University of Hong Kong, as the university awarded an honorary degree to Tsui in 2007. (Courtesy of the University of Hong Kong.)

Photo 6.9. Laureate Tsui receiving the degree of Doctor of Science, *honoris causa*, from The Chinese University of Hong Kong during its 55th Congregation in 1999. (Courtesy of the Chinese University of Hong Kong.)

Photo 6.10. Laureate Tsui with Professor Qifeng Zhou, President of Peking University, as the University's Honorary Doctoral Degree was conferred upon him in 2012. (Courtesy of Peking University.)

Photo 6.11. Laureate Tsui giving a seminar at Peking University in 2012. (Courtesy of Peking University.)

Photo 6.12. The book "*The Joy of the Search for Knowledge: A Tribute to Professor Dan Tsui*" *was* written to honor Laureate Tsui. (Courtesy of the World Scientific Publishing Co.)

"Soul Mountain", authored by Gao Xingjian in Chinese and translated into English by Mabel Lee. (Photo credit: Book cover from SOUL MOUNTAIN by GAO XINGJIAN and TRANSLATED BY MABEL LEE. Copyright © 2000 by Gao Xingjian. English language translation copyright © 2000 by Mabel Lee. Reprinted by permission of HarperCollins Publishers.)

Chapter 7

Xingjian Gao, 高行健
Literature, 2000

Mabel Lee

Nobel Laureate, Xingjian Gao, presents in Madrid his first Spanish-translated novel, "La montana del Alma" on May 16, 2001. (Courtesy of MARC ALEX/AFP/Getty Images.)

When Gao Xingjian won the Nobel Prize for Literature in 2000, it was the first time in the 100-year history of the prize that it had been awarded for a body of works originally created in the Chinese language. However, because Gao had become a French citizen in 1997, his win could not be regarded as a national victory for China, and the usual criticism of the choice of laureate followed. China watchers all over the world had been intent on picking the winner for more than two decades, and potential winners were constantly in the media spotlight. As Gao was not regarded as a potential Nobel winner, China watchers and experts had no inkling of his post-1987 publications or other creative activities.

Gao's Nobel award was based on the translations of his major works — two novels, collections of short stories and plays — into Swedish, French and English. His translators were established academics from various parts of the world, who after meeting Gao Xingjian, or reading his works had asked to translate his works. Göran Malmqvist (Swedish); Noël and Liliane Dutrait (French); Gilbert Fong (English); and Mabel Lee (English), not only translated his works, but also found the publishers. They and a small group of academics such as Henry Y. H. Zhao were also publishing analytical studies of Gao's works. By the end of 2000, Zhao had published Chinese and English editions of his monograph *Towards a Modern Zen Theatre: Gao Xingjian and Chinese Theatre Experimentalism* (1999 and 2000). Important also was the fact that when Gao received his award, his plays had performed on five continents, in different cultural settings and languages.

As a high-school student Gao had thought to pursue a career in art, but it was soon clear that this would mean painting propaganda posters for life. He instead enrolled in a five-year French studies course at the Foreign Languages Institute in Beijing, and being a voracious reader, he read through the library's entire holdings during this period. An intense impulse to write was kindled, but what he wanted to write was clearly contrary to the prescribed guidelines for writing socialist literature, so he wrote in secret. He graduated in 1962, and was allocated work as a translator and editor for the Foreign Languages Press in Beijing. At the beginning of the Cultural Revolution (1966–1976), he burned a suitcase of his unpublished plays, poems, fiction and notes. To have such writings discovered by the Red Guards would certainly have seen him severely punished.

He was over 40 years of age when he began his short literary career in China as a knowledgeable writer and playwright after the death of Mao Zedong and the end of the Cultural Revolution. His book *Preliminary Explorations into the Art of Modern Fiction* (1982) was widely read and acclaimed by veteran writers, academics, and university students. He also began publishing short stories and plays that were unlike anything others were writing: his main interest was exploring various narrative techniques to achieve convincing portrayals of the human psyche. With the support of veteran writers such as Ba Jin, he was reassigned work as a playwright for the People's Art Theatre, Beijing. His plays *Absolute*

Signal (1982), *Bus Stop* (1983), and *Wild Man* (1985) established his reputation in the Chinese theater world, as well as provided him with international credentials for being an experimental playwright with a good understanding of modern European literature. His early publications include *The Agony of Postmodernist French Literature* and *Prévert: French Modernist People's Poet*, as well as essays on Beckett, Artaud, Sartre, Camus, and the Polish playwrights, Grotowski and Kantor. His translations include Jacques Prévert's *Paroles* (1984) and Ionesco's *La Cantatrice Chauve* (1985).

Those in charge of safeguarding the "correct" path of the nation's cultural development were critical of the ambiguities present in Gao's writings, and he was singled out for various forms of harassment. His play *The Other Shore* (1986) was stopped at rehearsal, and without his knowledge, major publishers were quietly shelving his manuscripts. He had the opportunity to travel to Europe in late 1987, and while there applied for French residency to give himself some respite from China's politics. However, suddenly, Beijing student protesters in Tiananmen Square became a daily feature of TV news worldwide, and this continued until the military crackdown on June 4, 1989. This was a turning point in Gao's life. He ended his novel *Soul Mountain* that he had started writing in Beijing in 1982, and sent it off to his publisher in Taipei, and he quickly completed a play titled *Escape* (1989), a deeply meditative study of human psychology in a crisis situation, set in an allegedly anonymous square that is unmistakably Tiananmen Square. The play performed to appreciative European audiences, but annoyed the Chinese Democracy Movement because it had failed to portray the students as heroes. On the other hand, the Beijing authorities attacked the play for the sexual promiscuity of the three characters of the play, especially that of the woman.

Gao is certainly not alone in being genuinely interested in the accurate portrayal of female–male psychology that leads to the instigation of sexual relationships, and the continuation or termination of such relationships. It is the truth of this multi-layered existential reality that intrigues, and allows literary works to penetrate through temporal and spatial, as well as cultural and linguistic borders. However, he refused to conform to the dictates of any politics, including gender politics. After *Escape*, he wrote in succession three other plays about modern day female–male relationships: *Between Life and Death* (1991), *Dialogue and Rebuttal* (1992), and *Nocturnal Wanderer* (1993). His play, *Wild Man*, tells about Chinese village life in the 1980s. He notes a line chanted by the shaman on the creation myth of the Han Chinese people in *Record of Darkness*: in those times there were no rules for sexual liaisons, and people only knew their mothers....

Gao's writings inevitably evoke sympathy for women in their relationships with men, and speak of sexual urges and lust as a normal part of female behavior. He completed the first draft of his play *City of the Dead* in Beijing in 1987, and later in Paris produced the second and final drafts in 1990 and 1991. The play was first performed by the Hong Kong Dance Company in 1988. More than a decade later, based on the translation of Sookyung

Oh, who has translated a large number of his plays into Korean, a small theater production enthralled the audience in June 2011 as part of the Gao Xingjian Theatre Festival in Seoul. This was followed by 14 capacity performances of the play in February 2012 in the 540-seat Daehangno Arts Theatre in Seoul.

Throughout the 1990s, he continued to write at a feverish pace. His major works include the play *Of Mountains and Seas: A Tragicomedy of the Gods in Three Acts* (1993); the collection of plays and other performance pieces titled, *Weekend Quartet* (1996); the Peking Opera spectacular *Snow in August* (2000); the collection of critical essays *Without Isms* (1996); and his second autobiographical novel *One Man's Bible* (1999).

Delighted by the news of his Nobel win, he was totally unprepared for the media attack on his privacy. He was at the time already engrossed in planning the grand-scale "total theatre" production of *Snow in August* scheduled to take place 19–22 December 2002 in the National Opera House, Taipei. He would retrain traditional Peking Opera performers, and involve himself in all other aspects of the production, including the choreography. The music was in the process of being finalized by composer Xu Shuya who had relocated to Paris in 1988 from Shanghai, and was distinguishing himself at the highest levels of his profession. Gao's own artworks would form the backdrops for the performance, and "total theatre" meant that the action, music, acrobatics, martial arts, and the stunts of traditional Peking Opera and local opera would also be introduced. The production would enlist a choir and a symphony orchestra amounting to a total of 100 performers.

It was while rehearsing for this highly ambitious production during late 2002 that Gao Xingjian had his first serious brush with death. He recovered sufficiently to direct (with assistance) the premiere performance of *Snow in August* before rushing back to Paris to direct the Comédie Français production of his play *Weekend Quartet*, a play that was one of several that he had written first in French and then in Chinese. In February and March of 2003, he underwent heart surgery. The year 2003 had been designated "Gao Xingjian Year" by the City of Marseille during which Gao would direct *Snow in August* at Opéra de Marseille, and his new play *Le Quêteur de la Mort* at Théâtre du Gymnase. However, it was during rehearsals for the latter that he collapsed again, and the play was staged with the help of co-director Romain Bonnin, 23–26 September, 2003. Large exhibitions of his artworks had been held in Marseille earlier that year, but the performance of *Snow in August* was postponed. This time his physical collapse incapacitated him, and his doctors warned that medications and a total change of lifestyle were needed. He himself was acutely aware that he could not read anything without sending his blood pressure soaring, and inducing heart palpitations.

It took much of 2004 to recuperate, and during this time, it was only through painting, poetry and thinking about filmmaking that he was able to gratify his aesthetic and intellectual curiosity. He had depended on painting to sustain his literary endeavors and everyday life since relocating to Paris, and his most influential exhibitions include

"La Fin du Monde" at Ludwig Museum, Germany (2007) and "Depois do dilúvio" at Sintra Museum of Modern Art, Portugal (2009). However, it is Belgium, in the city of Brussels, that Gao Xingjian's art has been honored in a very exceptional manner. In late February 2015, a three-month retrospective of his paintings opened at Ixelles Museum on the day after his work *The Awakening of the Consciousness* went on permanent exhibition in a large dedicated room located on the ground floor of the Royal Museums of Fine Arts of Belgium. In the six large-scale works comprising *The Awakening of the Consciousness*, Gao explores intangible psychological states designated as *Subconscious, Illusion, Impulse, Introspection, Somewhere Else,* and *Bewilderment.*

Silhouette/Shadow (2007), *After the Flood* (2008) and *Requiem for Beauty* (2013), are the three films that he has completed to date, and each incorporates a large range of painting, photography, dance, and sound possibilities. Marketing these innovative art films proved to be impossible, but they are shown at his art exhibitions or attached to his art books. His first collection of poems *Wandering Spirit and Metaphysical Thoughts* (2012) reflects his deep concerns for we humans who live in the present times, and informed by black humor and absurdist aesthetics, these uniquely playful poems resonate loudly with truth, indicating that 2000 Nobel Laureate Gao Xingjian is satisfying his ever-surging flood of creative impulses.

Selected Books in English for Further Reading
Translations from the Chinese unless otherwise indicated

BERGEZ, Daniel. *Gao Xingjian: Painter of the Soul*. London: Asia Ink, 2013.

GAO, Xingjian. *The Other Shore: Plays by Gao Xingjian*. Trans. Gilbert C. F. Fong. Hong Kong: The Chinese University Press, 1999.

—————. *Soul Mountain*. Trans. Mabel Lee. Sydney, New York, London: HarperCollins, 2000.

—————. *Return to Painting*. Trans. Nadia Benabid from the French. New York: HarperCollins, 2001.

—————. *One Man's Bible*. Trans. Mabel Lee. Sydney, New York, London: HarperCollins, 2002.

—————. *Snow in August*. Trans. Gilbert C. F. Fong. Hong Kong: The Chinese University Press, 2003.

—————. *Buying a Fishing Rod for My Grandfather*. Trans. Mabel Lee. Sydney, New York, London: HarperCollins, 2004.

—————. *The Case for Literature*. Trans. Mabel Lee. Sydney: HarperCollins, 2006; New Haven and London: Yale University Press, 2007.

—————. *Escape & The Man Who Questions Death*. Hong Kong: The Chinese University

Press, 2007.

———. *La Fin du Monde*. Bielefeld: Kerber Verlag, 2007.

———. *Of Mountains and Seas: A Tragicomedy of the Gods in Three Acts*. Trans. Gilbert C. F. Fong. Hong Kong: The Chinese University Press, 2008.

———. *Depois do dilúvio*. Barcelona: El Cobra Ediciones, 2009.

———. *Aesthetics and Creation*. Trans. Mabel Lee. Amherst NY: Cambria Press, 2012.

———. *City of the Dead* & *Song of the Night*. Trans. Gilbert C. F. Fong and Mabel Lee (respectively). Hong Kong: The Chinese University Press, 2015.

———. *After the Flood* (photography, incl. CD). Taipei: Lianjing, 2015.

———. *Requiem for Beauty* (photography, incl. CD). Taipei: Lianjing, 2016.

KUO, Jason C. *The Inner Landscape: The Paintings of Gao Xingjian*. Washington DC: New Academia, 2013.

ŁABĘDZKA, Izabella. *Gao Xingjian's Idea of Theatre: From the Word to the Image*. Leiden and Boston: Brill, 2008.

LACKNER, Michael, and Nikola Chardonnens, eds. *Polyphony Embodied: Freedom and Fate in Gao Xingjian's Writings*. Berlin and Boston: Walter de Gruyter GMbH, 2014.

QUAH, Sy Ren. *Gao Xingjian and Transcultural Chinese Theater*. Honolulu: University of Hawai'i Press, 2004.

SZE-LORRAIN, Fiona, ed. *Silhouette/Shadow: The Cinematic Art of Gao Xingjian*. Paris: Contours, 2007.

TAM, Kwok-kan, ed. *Soul of Chaos: Critical Perspectives on Gao Xingjian*. Hong Kong: The Chinese University Press, 2001.

YEUNG, Jessica. *Ink Dances in Limbo: Gao Xingjian's Writing as Cultural Translation*. Hong Kong: Hong Kong University Press, 2008.

ZHAO, Henry Y. H. *Towards a Modern Zen Theatre: Gao Xingjian and Chinese Theatre Experimentalism*. London: School of Oriental and African Studies, 2000.

Editors' Note
Honors (Wikipedia)

- 1992, *Chevalier de l'Ordre des Arts et des Letters*
- 2000, Premio Letterario Feronia in Rome
- 2001, Honorary doctorate by *Chinese University of Hong Kong*
- 2001, Honorary doctorate by *National Sun Yat-sen University*
- 2002, Honorary doctorate by *National Chiao Tung University*
- 2002, *Legion of Honour* by then French President *Jacques Chirac*
- 2003, *l'Anne Gao Xingjian*, the City of *Marseille*
- 2005, Honorary doctorate by *National Taiwan University*

Photo 7.1. Xingjian Gao receiving the Nobel Prize in Literature from King Carl XVI Gustaf of Sweden. Stockholm in 2000. (Photo credit: Jonas Ekström/Scanpix Sweden/Sipa USA.)

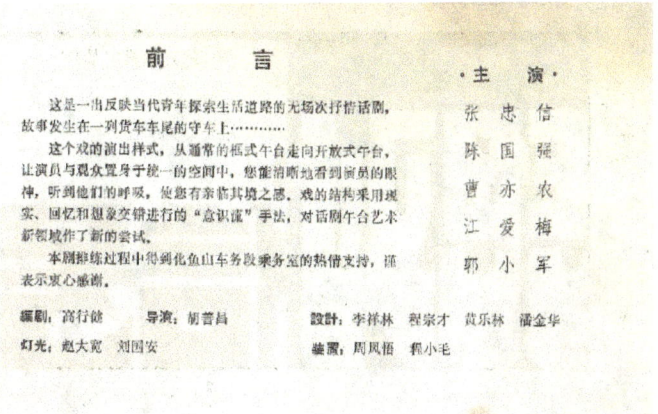

Photo 7.2. A vintage photograph of an early 1980s ticket for Gao Xingjian's first play production in China: Absolute Signal.

Photo 7.3. Front cover of a well-received book of essays penned by Gao Xingjian and translated by Mabel Lee, *Aesthetics and Creation*. Amherst, NY: Cambria Press, 2012. (Reproduced with permission © 2012 by Cambria Press.)

Photo 7.4. Gao Xingjian receiving a Doctorate in Literature, *honoris causa*, at The Chinese University of Hong Kong in 2001. (Photograph, courtesy of The Chinese University of Hong Kong.)

Photo 7.5. Editions L'Aube first introduced readers to French translations of Gao Xingjian's writings in the 1990s, and on 18 March 2004, was ready to promote his works in Paris again. (Photograph courtesy of MEHDI FEDOUACH/AFP/Getty Images.)

Photo 7.6. Gao Xingjian receives a Doctorate, *honoris causa*, at the ceremony of the 175th Anniversary of the University of Brussels, 5 May, 2010. (Photograph courtesy of JULIEN WARNAND/AFP/Getty Images.)

Photo 7.7. During a whirlwind HarperCollins USA promotion tour of Soul Mountain in 2001, Gao Xingjian was delighted to meet with 1976 Nobel Laureate Saul Bellow in New York. (Photograph, courtesy of Mabel Lee, Sydney.)

Photo 7.8. Gao Xingjian and his English-language translators Gilbert C. F. Fong and Mabel Lee in Stockholm for the Nobel ceremonies in December 2000. (Photograph, courtesy of Gilbert C. F. Fong, Hong Kong.)

Photo 7.9. In early July 2000, Gao Xingjian visited Sydney to launch the English edition of his *Soul Mountain*. (Photograph, courtesy of Mabel Lee, Sydney.)

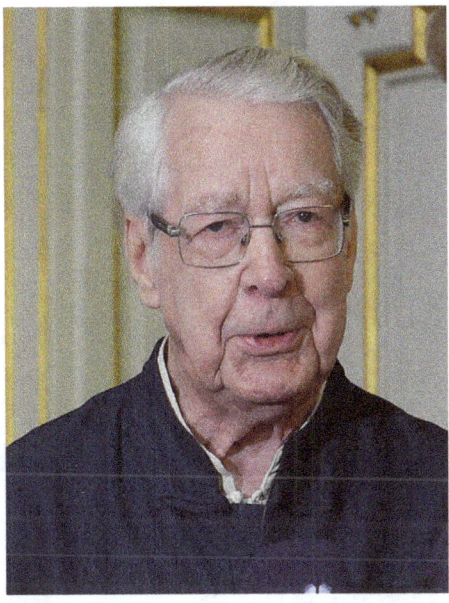

Photo 7.10. Eminent sinologist and linguist, Göran Malmqvist translated Gao Xingjian's *Soul Mountain* from Chinese into Swedish in 1992. (Photo credited to Frankie Fouganthin — Own work. Created 11 October 2012. (From: Wikimedia Commons.)

Photo 7.11. Gao Xingjian translator Mabel Lee enjoying a chat with Gao Xingjian's novelist wife, Céline Yang. They are standing in front of a large Gao Xingjian artwork at his iPreciation Gallery exhibition in Singapore, November 2013. (Photograph, courtesy of iPreciation Gallery, Cuscaden Street, Singapore.)

Photo 7.12. Academic Noël Dutrait and his wife, Liliane (passed away on 4 October 2010), worked as a team to produce French translations of many of Gao Xingjian's major works. Noël is photographed here in conversation with Gao during a literary festival event in 2011. (Photograph, courtesy of Noël Dutrait.)

"It is difficult to say what is impossible, for the dream of yesterday is the hope of today and the reality of tomorrow."
Robert H. Goddard.

(A note from an early page of "Replacement of Renal Function by Dialysis". In: William Drukker, Frank M. Parsons, John F. Maher (eds), 2nd edn., Martinus Nijhoff Publishers, Boston, 1983.)

Chapter 8

Roger Yonchien Tsien, 錢永健
2008 Nobel Laureate in Chemistry
2004 Wolf Laureate in Medicine

Keith K. Lau, Mohamed A. Rahman, Yuzhu Bian, Yun Wang and Thomas M. S. Chang

2008 Nobel Laureate in Chemistry. (Photo courtesy of Laureate Tsien.)

Nobel Prize motivation for Roger Y. Tsien (from Nobelprize.org): *"for the discovery and development of the green fluorescent protein, GFP"* (Roger Y. Tsien — Nobelprize.org.)

American Chinese biochemist Roger Yonchien Tsien has revolutionized the fields of cell biology and neurobiology by formulating ways which allow scientists to use novel fluorescent proteins to track the behavior of intracellular molecules in real time. For this scientific contribution, Roger was awarded the Wolf Prize in Medicine in 2004. Roger also shared the glory of the Nobel Prize in Chemistry in 2008 with two other distinguished scientists: Osamu Shimomura (下村修) from Boston University and Martin Chalfie from Columbia University. These three scientists were selected to receive the award for their works on the green fluorescent protein (綠色螢光蛋白質).

Roger was born on February 1, 1952 in New York City to Hsue-Chu Tsien (錢學榘) and Yi-Ying Li (李懿穎). Hsue-Chu was an engineer and Yi-Ying, a nurse. Hsue-Chu received the Gengzi Indemnity Scholarship (also called the Boxer Indemnity Scholarship; 庚子賠款獎學金) and studied aeronautical engineering at the Massachusetts Institute of Technology. He also served as the Deputy Chief Engineer of the National Air Aviation Commission in China during World War II, and later relocated back to America at the end of the war, where he served as a consultant at the Esso and Boeing Companies for years before his retirement. Roger's family originated from Hangzhou (杭州市), Zhejiang Province (浙江省), China.

Tracing back his genealogy, he is the 34th generational descendant of Emperor Liu Tsien (錢繆) of the Wuyue (吳越) Kingdom. Tsien's relatives are very accomplished. His father's cousin, the late rocket scientist Hsue-shen Tsien (錢學森), was known as one of the founders of the Jet Propulsion Laboratory of the California Institute of Technology and as the director of various Chinese ballistic-missile and space programs.

Roger grew up in Livingston, New Jersey, and attended Livingston High School. In his Nobel biography, Roger recalled that his family sometimes encountered racial prejudice. When they applied to buy a house in Livingston, New Jersey, the developer refused to sell to them, saying that they could not permit Livingston to become a Chinatown, nor could they afford the likelihood that other customers would refuse to buy houses next to a Chinese family. Little did they know that the children of this family all grew up to become leaders in academic and professional worlds. Roger's immense interest in chemistry stemmed from a young age. He has always been fascinated by how chemicals changed colors during serial reactions. He fought going to kindergarten until his teacher allowed him to bring in a favorite book: he picked "All about the Wonder of Chemistry". From the age of eight, he performed increasingly complex and sometimes hazardous chemistry experiments at home.

At the age of 16, he won first prize in the prestigious nationwide Westinghouse Science Talent Search (now called the Intel Science Talent Search), a research-based competition in science for high school students. Even former President George H. W. Bush acknowledged the competition as the Super Bowl of Science. In April 1968, he had to choose among four colleges: Columbia, MIT, Caltech, and Harvard. Roger eventually decided to attend Harvard University on a National Merit Scholarship. The diversity of courses at Harvard let him sample art history, visual design, economics, music theory and chamber music performance in addition to chemistry and physics. However, he chose to major in neurobiology and graduated with a Bachelor of Science in Chemistry and Physics in 1972.

After graduation, he felt it was time to broaden his horizon, so he applied for a Marshall Scholarship to go to the other Cambridge. Roger's brother Richard, an accomplished neurobiologist and member of the United States National Academy of Sciences, was a Rhodes Scholar and was at Oxford from 1966 to 1970. Roger spent nine years at the Physiology Laboratory at the University of Cambridge. First, he was a PhD student at Churchill College with the eminent muscle physiologist, Richard Adrian; then he did a postdoc with Timothy Rink and was a research fellow at Gonville and Caius College, Cambridge. It was during this time that he met his future wife, Wendy Marchant Globe, at a Christmas party in 1976 when he was working with Tim Rink.

It is interesting to note that Roger's postgraduate training was in physiology, rather than in chemistry. To study neuronal activity, it became necessary to find ways to synthesize dyes to track the electrical activities of the nerve cells. Much of Roger's early work was directed at imaging neural activity, by trying to develop tracers of sodium- or calcium-ion movement that support brain signaling. He met Ian Baxter at that time and started to learn how to synthesize organic materials that will allow the study of biological functions. He saw the necessity of developing a proper chemical framework for probes. By 1980, he had invented quin-2, a synthetic fluorescent dye that selectively binds to calcium, and had devised a clever way to sneak this dye and other probes into intact cells. This first practical probe for calcium found early use in studies of intracellular cell signaling. This experience sparked Roger's research direction for decades to come.

In 1982, Roger joined the physiology department at the University of California, Berkeley. Roger was then well known for his insights into the design of molecular probes and the use of fluorescence. He was encouraged by his colleagues to create more tools and better molecular probes to monitor intracellular free calcium, intracellular pH and sodium activity. Roger was generous in providing materials to other scientists and in helping his colleagues and others in developing this new technology.

Facing resource constraints, Roger transferred to UCSD in 1989, where he remained for the rest of his life. His main focus was to make sensors that could be genetically encoded, allowing scientists to target specific cell types without having to inject a tracer into the cell. It was at this time that he saw the potential of the jellyfish green fluorescent protein (GFP). The protein had been isolated from jellyfish in the 1960s by Osamu Shimomura and cloned by Douglas Prasher in 1992. Martin Chalfie, who also shared the Nobel, first used GFP to image living cells in 1994. Roger, however, pioneered the development of GFP variants. Through a combination of rational design and random mutagenesis, he created a dozen of bright fluorescent proteins of various colors based on GFP. He is responsible for much of our understanding of how GFP works and for developing new techniques and mutants of GFP.

The Swedish Academy described GFP as "a guiding star for biochemists, biologists, medical scientists and other researchers." The GFP is different from luciferin (螢光素), which is another light-emitting compound. Luciferin was first isolated from fireflies (螢火蟲) and requires the enzyme luciferase in order to emit the fluorescent light. In contrast, the GFP does not require any catalysts to generate bioluminescence. In 1962, Frank Johnson and Shimomura (the other Nobel Prize laureate who shared the 2008 Chemistry Prize with Tsien) studied the bioluminescense of *Aequorea victoria*, a jellyfish found abundantly in Northern Pacific Ocean. Johnson and Shimomura isolated the two compounds responsible for the specific fluorescent color of jellyfish: the aequorin (水母素) and the GFP.

Underlying the search for the GFP is a theme that is prominent in the field of science: besides having vision, hard work and perseverance, luck can sometimes play a pivotal role in the success of a scientist. Just as Louis Pasteur said more than a hundred years ago that chances favored a prepared mind. Initially, the clinical application of this fluorescent protein was not appreciated, even after the gene that encodes for the green fluorescence protein had been cloned and the cDNA, synthesized by Douglas Prasher in 1990. Prasher gave the cDNA to Martin Chalfie and Roger, and then after a series of mishaps, left the academic field. Chalfie further pursued this line of research and was eventually able to express the GFP in other organisms such as *Escherichia coli* (大腸桿菌), a bacterium found usually in the human bowel, and in *Caenorhabditis elegans* (秀麗隱桿線蟲), a round worm. Chalfie's research provided the first evidence that the green fluorescent protein was unique since it did not require the presence of any exogenous substance or cofactor for the emission of light. It was at this point that the scientific community finally saw and acknowledged the significance of the GFP.

In the mean time, Roger was investigating the structure of the GFP and the mechanisms behind its fluorescent properties. After studying the protein's structure, Roger was able to modify and synthesize derivatives possessing improved fluorescence intensity and stability. He also found that oxygen, which is abundant inside living cells, was the only factor necessary for activating fluorescence from the protein. The protein was found to be non-toxic when expressed under most experimental conditions and could be bound to other proteins without altering or perturbing each other's functions within a cell. By artificially modifying the gene responsible for the protein, Roger created a spectrum of fluorescent colors, such as tdTomato and mCherry (which are orange-red colors). This approach enabled scientists to label various intracellular proteins and study their different behaviors at the same time. GFP variants are now ubiquitous in biological research. These unique properties render the GFP and its derivatives as being exceptionally suitable in many streams of research in cell biology. One exciting clinical application of the GFP involves the specific expression of fluorescent proteins in tumor cells. Roger openly admitted that because both his father and his mentor, Professor Richard Adrian, had died of cancer, he was very devoted to relating his research to cancer studies. Using the mouse model, Roger's group attempted to develop innovative ways to guide cancer therapy by labelling tumor cells with fluorescent proteins. Specifically, the design of fluorescent tracer to illuminate tumor during cancer surgery, and also the fundamental study to understand the storage of long-term memory in the brain.

Roger had a passion for adding colors to his research. He once told his students that it is essential to put some pleasure at work in order to make it sustainable. People tend to believe in what they see; however, even the best microscopes are not capable of seeing the activities or tracking the motions of molecules inside cells. But, it was the discovery of the GFP that made all these activities visible. Based on the GFP, Roger further designed and synthesized other fluorescent proteins that are capable of tracking signal transduction inside living cells by lighting up intracellular molecules with color, thus allowing the visualization of the movement of the molecules in real time. An illustrative example is the export of genetic materials from the nucleus of a cell. Roger likened his experience with fluorescent proteins during experiments to being able to communicate with the cells — as if the cells were alive and talking back to him!

In his own biography, Roger mentioned that his career has been "shaped by a strange mixture of chances and fateful predisposition". Besides having a mentor who had "re-instilled" his enjoyment in chemistry, Roger was also very thankful to Dr. Douglas Prasher, who had cloned the gene for the GFP and shared with him the cDNA without asking for anything in return. Roger wondered why he had been honored and not Prasher, who had shared the gene with both Chalfie and him. When Roger was interviewed by the Nobel Committee on Chemistry in 2004, he recommended Prasher and Shimomura to be the Nobel winners based on his belief that the contributions of these two scientists were foundational to the development of the GFP.

Dr. Prasher had by then dropped out of science. But Chalfie and Roger invited him to attend the Nobel ceremony in Stockholm and paid for his trip. In 2012, Roger hired Dr. Prasher to work in his lab.

Amongst his numerous academic honors, besides the Wolf Prize in Medicine (shared with Robert Weinberg, 2004) and the Nobel Prize in Chemistry (2008), others particularly worth mentioning include the Searle Scholar Award (1983), the Artois-Baillet-Latour Health Prize (1995), the Gairdner Foundation International Award (1995), the Award for Creative Invention from the American Chemical Society (2002), the Heineken Prize in Biochemistry and Biophysics (2002), the Rosenstiel Award (2006), and the E.B. Wilson Medal of the American Society for Cell Biology (shared with M. Chalfie, 2008). Moreover, Roger is a member of the American Academy of Arts and Sciences, the National Academy of Sciences, the Royal Society of London, and the Academia Sinica in Taiwan. To capitalize on his inventions, Roger co-founded two companies in the biotechnological industry. The Aurora Biosciences Corporation commercialized drug discovery tools using fluorescent markers, and the Senomyx company looked for ways to modulate taste receptors to reduce the amount of sugar and salt in food without affecting taste.

Editors' Note

It is worth noting that Laureate Tsien received, from around the world, myriad honors and awards, accolades which he richly deserved.

September 1, 2016. We deeply mourn the passing of our brilliant and visionary Laureate Roger Tsien on August 24, 2016. He will be remembered not only for his great intellect, but also for his sense of humility and humor so rare in someone of his caliber. For many of us, we will never forget his generosity in sharing his molecular probes and for coming to the Chinese American Society of Nephrology Annual Meeting in 2003 to deliver a distinguished scientist lecture to a small group of Chinese-American nephrologists and young investigators. Roger was full of warmth, humor, and of course color. His lasting legacy is the simplicity and the beauty of the tools that he developed. As Roger's wife Wendy said, "Roger was ahead of us all. He was ever the adventurer, the pathfinder, the free and soaring spirit. Courage, determination, creativity, and resourcefulness were hallmarks of his character. He accomplished much. He will not be forgotten". We will all miss him dearly.

References and Suggested Readings

1. Roger Y. Tsien — Facts — Nobelprize.org. http://www.nobelprize.org/nobel_prizes/chemistry/laureates/2008/tsien-facts.html. Accessed on June 20, 2015.

2. Roger Y. Tsien — Biographical — Nobel Prize. http://www.nobelprize.org/nobel_prizes/chemistry/laureates/2008/tsien-bio.html. Accessed on March 31, 2015.
3. Tsien RY. Breeding molecules to spy on cells. *Harvey Lecture* 2003–2004; **99**: 77–93.
4. Roger Tsien. (2007) Bringing color to cell biology. *J Cell Biol* **179**(1): 6–8.
5. Tsien RY. (2009) Constructing and exploiting the fluorescent protein paint box (Nobel Lecture). *Angew Chem Int Ed Engl* **48**(31): 5612–26.
6. Tsien RY. (2009) Indicators based on fluorescence resonance energy transfer (FRET). Cold Spring Harb Protoc. 2009(7): pdb.top57. doi: 10.1101/pdb.top57.
7. Tsien RY. (2010) The 2009 Lindau Nobel Laureate Meeting: Roger Y. Tsien, Chemistry 2008. *J Vis Exp* (**35**). p. ii: 1575. doi: 10.3791/1575.
8. Nguyen QT, Olson ES, Aguilera TA, Jiang T, Scadeng M, Ellies LG, Tsien RY. (2010) Surgery with molecular fluorescence imaging using activatable cell-penetrating peptides decreases residual cancer and improves survival. *Proc Natl Acad Sci USA* **107**(9): 4317–22.
9. Shu X, Lev-Ram V, Deerinck TJ, Qi Y, Ramko EB, Davidson MW, Jin Y, Ellisman MH, Tsien RY. (2011) A genetically encoded tag for correlated light and electron microscopy of intact cells, tissues, and organisms. *PLoS Biol* **9**(4): e1001041. doi: 10.1371/journal.pbio.1001041.
10. Roger Y. Tsien — Wikipedia, the free encyclopedia. http://en.wikipedia.org/wiki/Roger_Y._Tsien. Accessed on March 31, 2015.
11. Tsien lab Website. http://www.tsienlab.ucsd.edu. Accessed on March 31, 2015.
12. Douglas Prasher — Wikipedia, the free encyclopedia. http://en.wikipedia.org/wiki/Douglas_Prasher. Accessed on March 31, 2015.
13. Tsien R. Y. (2013) Very long-term memories may be stored in the pattern of holes in the perineuronal net. *Proc Natl Acad Sci USA* **110**(30): 12456–61.
14. Jorgenson L. A., Newsome W. T., Anderson D. J. et al. (2015) The Brain Initiative: developing technology to catalyse neuroscience discovery. *Phil Trans R Soc Lond B Biol Sci* B370 (1868). Phil: 20140164.
15. "Roger Y. Tsien, chemist shared Nobel for tool to research Alzheimer's, dies at 64". The Washington Post. 31 August 2016.
16. Roger Y. Tsien (1952–2016): Nature: Nature Research www.nature.com/nature/journal/v538/n7624/full/538172a.html by T. J. Rink — 2016.
17. Douglas Prasher from The San Diego Union-Tribune. http://www.sandiegounion tribune.com/lifestyle/people/sdutcientist-turned-shuttle-van-driver-2013apr13-story, amp.html).
18. Rink TJ, Tsien LY, Tsien RW. (2016) Roger Yonchien Tsien (1952–2016). *Nature* **538**(7624): 172. doi:10.1038/538172a.

Photo 8.1. Roger Tsien with his family. From left to right: his brothers, Richard Tsien and Louis Tsien; his father, Hsue-Chu Tsien; Roger (8-years old); his mother, Yi-Ying Li. (Copyright © The Nobel Foundation.)

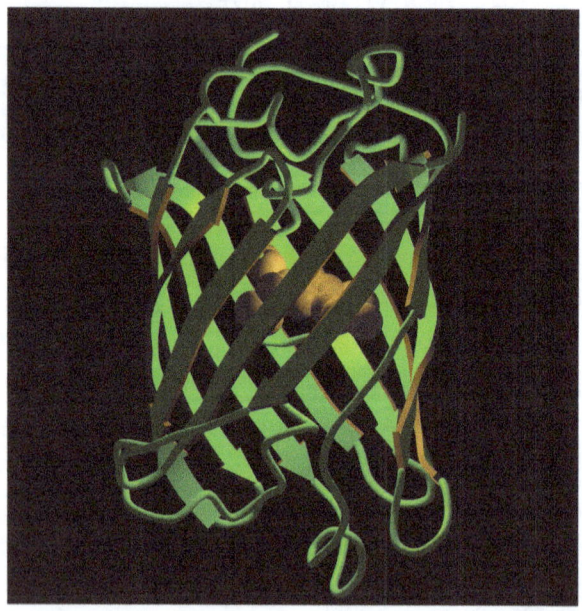

Photo 8.2. The green fluorescent protein, GFP, is shaped like a cylinder. (Courtesy of Laureate Tsien.)

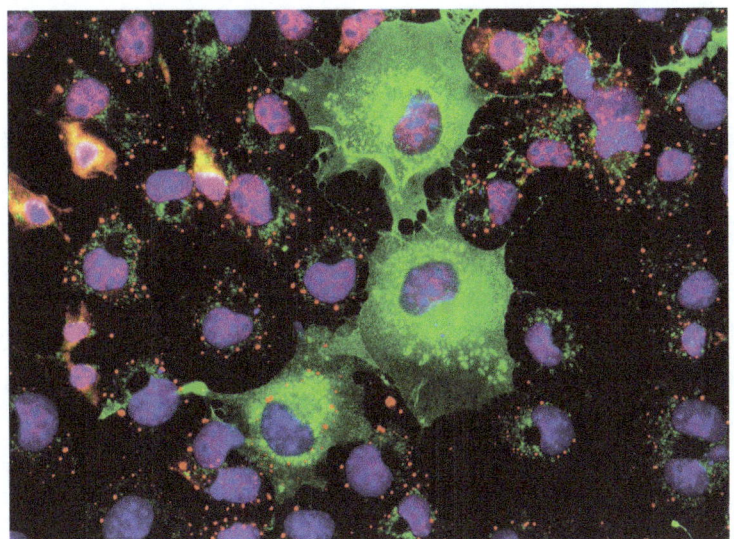

Photo 8.3. Laureate Tsien has made important contributions to the fluorescence technique. The above picture is an example of the use of the fluorescence approach to track proteins, showing cancer cells (shown in green) adapting to starvation. (Courtesy of Nancy Kedersa/Science Fraction/Getty Images.)

Editors' Note: This figure exemplifies the immense value of the fluorescence method by which components of structures can be identified from differences in coloring. For instance, abnormalities in cancer cells can be detected in this way.

Photo 8.4. Roger Tsien receiving the Nobel Prize from Carl XVI Gustaf, the King of Sweden in 2008. (Courtesy of Fredrik Sandberg/Scanpix/Sipa USA.)

Photo 8.5. Laureate Tsien after having received his Nobel Prize at the Stockholm Concert Hall on December 10, 2008. (Copyright © The Nobel Foundation 2008, Photo: Hans Mehlin.)

Photo 8.6. Laureate Tsien (L1), Laureate Osamu Shimomura (L2), Laureate Martin Chalfie (L4) and fellow 2008 Nobel laureates. (Copyright © The Nobel Foundation 2008, Photo: Mia Åkermark Orasis.)

Photo 8.7. Laureate Tsien presenting a talk on the topic of Green Fluorescent Protein during a visit to the Värmdö Gymnasium School in Stockholm on December 12, 2008. (Copyright © Nobel Media AB, Photo: Karin Svanholm.)

Photo 8.8. Roger Tsien and his wife, Wendy Marchant Globe, in Stockholm around the time of the Nobel award ceremony in 2008. (Courtesy of Scanpix Sweden/Sipa Press.)

Photo 8.9. President George W. Bush celebrating, with Laureate Roger Tsien (R1), Laureate Martin Chalfie (R2), a co-winner of the Nobel Prize in Chemistry, and Laureate Paul Krugman (L1), the winner of the Nobel Prize in Economics, in the Oval Office at the White House on November 24, 2008 in Washington, DC. (Courtesy of Chip Somodevilla/Getty Images News/Getty Images.)

Photo 8.10. Receiving the degree of Doctor of Science, *honoris causa*, from the University of Hong Kong in 2009. (Courtesy of the University of Hong Kong.)

Photo 8.11. Receiving the degree of Doctor of Science, *honoris causa*, from the Chinese University of Hong Kong in 2009. (Courtesy of The Chinese University of Hong Kong.)

Photo 8.12. Biochemist Douglas Prasher in the University of California at San Diego laboratory led by Professor Roger Tsien [John Gastaldo]. (Courtesy of The San Diego Union-Tribune.)

The Chinese University of Hong Kong (CUHK) [with its beautiful, bucolic campus] where Laureate Charles Kao served as the Vice-Chancellor from 1987 to 1996. (Photo courtesy of CUHK.)

Chapter 9

Charles K. Kao, 高錕
Nobel Prize in Physics, 2009

Kenneth Young

A young Charles Kao. (Photo courtesy of Mrs. Gwen Kao.)

Charles Kuen Kao, 高錕[1] received the 2009 Nobel Prize in Physics for his "ground-breaking achievements concerning the transmission of light in fibers for optical communication"[2] — work done in the 1960s.

Early Life and Education

Charles Kao was born in 1933, in Shanghai, China, to a family with a scholarly tradition. His grandfather belonged to what would be known in the West as the landed gentry, and moved in literati circles. His father, Chun-Hsiang Kao, 高君湘, was a lawyer, having been educated at the University of Michigan in the US. In 1949, "ahead of the civil war front approaching Shanghai",[3] the family moved to Hong Kong; Charles was then 16. He attended St. Joseph's College, a secondary school run by the La Salle Brothers. English was used in the classroom (a practice that continues to this day) and on the playground. As a result, despite a number of teenage years in Hong Kong, Kao never became proficient in the local dialect, even though he spent a considerable part of his career and now his retirement in the city.

Upon completing secondary school, Kao went on to Woolwich Polytechnic (now Greenwich University) in the UK to complete his A-levels. To nobody's surprise, he "easily" obtained A's in four subjects: pure mathematics, applied mathematics, physics, and chemistry- A's were worth something in those days, before the blight of grade inflation, and four A's were exceptional. He would have been accepted to a wide choice of institutions, but he elected to stay at Woolwich Polytechnic because he had an enjoyable time there.[3] He obtained his BSc in Electrical Engineering in 1957, but his grades were not stellar, on account of a busy social life and a fondness for tennis (a sport he was able to continue into his late 70s).

After graduation, Kao joined Standard Telephones and Cables (STC) in their factory in North Woolwich nearby. There he met Gwen, a fellow engineer, and they were married in 1959. Soon after this, Kao accepted a teaching post at the then Loughborough College of Technology (later Loughborough University of Technology and still later Loughborough University). However, his resignation from STC was met with a counter offer to transfer to the company's Research Labs in Harlow. At that time, STC was a subsidiary of the US conglomerate International Telephone and Telegraph (ITT) and a major player in the telecommunications industry in the UK. Their research division, Standard Telecommunications Laboratories (STL), like other telecom labs at the time, was looking for ways to transmit large volumes of data over long distances — large and long by the standards of those days. Being presented with the problem was a stroke of good luck, and in this instance, fortune also shone on a mind that was not only prepared but way ahead of its time.

The Key Discovery

Signals are carried by electromagnetic waves. Distance by itself was not a problem: even at the beginning of the 20th century, Marconi was already able to send radio signals across the Atlantic, a distance of some 3,000 km. However, the volume of data was a problem. As the amplitude of the wave varies, one can code "0" for, say, low amplitude and, correspondingly, "1" for high amplitude.[4] The absolute theoretical limit to the transmission capacity per second is therefore proportional to the *frequency f* of the waves (technically the carrier frequency); that is, how many times the signal can go up and down per second.[5] Marconi's radio transmission operated at ~1 MHz; the frequency of microwaves can be a few hundred thousand (10^5) times higher, and that for light a few hundred million (10^8) times higher — if either could be harnessed for this purpose, a new age in communications could open up.

One obvious method was to transmit electromagnetic waves through the atmosphere. Lenses placed at suitable intervals ensure that the beam does not diverge too much; repeaters amplify the signal before sending it on to the next relay point.[6] The scattering of the beam by density fluctuations in the atmosphere would have to be dealt with. Thus, a much better method would be to confine the waves somehow.

Two approaches were studied. "The race between circular microwave waveguides and optical communication was on, with the odds heavily in favor of the former".[7] The latter method — much less obvious at the time — was to send the light down a thin glass fiber. Any ray of light that propagates in a direction close to the axis would strike the boundary at a glancing angle and be confined by internal reflection, thus zigzagging down the fiber without divergence. This seems obvious now to anyone who has seen the many ornamental displays made with such fibers. There is a technical problem: when the fiber diameter is not much larger than the wavelength, namely when it is on the micrometer scale, one really cannot speak of corpuscular rays, but must deal with the wave nature of light. Kao's groundbreaking 1966 paper, co-authored with George Hockham,[8] starts on its first page with an analysis of the waves using Bessel functions and Hankel functions, though that in itself was only an inconvenience, not an obstacle in any sense.

One well-known serious problem that glass fibers faced was so "elementary" and "obvious" without the need to invoke any fancy mathematics that everyone thought that glass fibers were doomed. A pane of glass a few centimeters thick may be transparent, but a slab a few meters thick would begin to look opaque. In other words, there is some small absorption of light that limits the distance L beyond which the intensity would show a significant attenuation.[9] Kao and Hockham thought that to make fibers feasible as a transmission medium, one needs to have $L > 500$ m,[10] but the best glass available at the time had $L \sim 50$ m;[11] there was a factor of 10 to be bridged. The key insight of Kao and Hockham[8] was that this level of absorption was not intrinsic, but mostly due to trace impurities, for example iron, at levels of parts per million, and that "the lower limit of loss imposed by fundamental mechanisms"[12] was actually very small. In other words, they pointed out that *glass can be made to be extremely transparent*. In many ways, this was the most important point of their paper.

It took much explanation, effort, and indeed much evangelizing by Kao against ingrained skepticism, before that insight started to sink in. Research and development on the physical properties and the production of high-quality glass gained momentum. For example, Corning, which had been known as a producer of kitchenware, was one of the somewhat unexpected and successful entrants into this line of work. Optical fibers were many orders more difficult — and also more profitable — than even the most expensive dishes intended for the oven. By 1984, attenuation lengths of $L \sim 20$ km were achieved.[12] The problem of transmission distance was basically solved.

The 1966 paper of Kao and Hockham[8] was surprising in another respect. Although the attenuation problem was acknowledged to be unsolved at that time, the authors nevertheless set that aside boldly and went on to consider a host of design and implementation issues. Many of the proposals were prescient and correctly predicted the path of future development; for example, single-mode operation was discussed.[13] Now the technology of choice for long-distance transmission and therefore often regarded as "natural", single-mode fibers was in fact far from obvious: the fiber would be made extremely thin, very little light would be fed into it, and the rays (to use a language that is somewhat heuristic when the fiber is so thin) would bounce from wall to wall in only one way, which might seem unnecessarily limiting. This is analogous to saying that a thin pipe would carry more water. But Kao and Hockham[8] correctly realized that it was precisely this limitation to only one zigzag path that prevents rays starting at the same time from the source from arriving at the other end out of synchronization, which would distort the signal.[14]

All these results were also contained in his thesis on "Quasi-optical waveguides," for which Kao was awarded a PhD by the University of London in 1965.

Other Enabling Technologies

It took decades for this deep insight to become reality. The necessary improvements in glass technology were one factor, but other enabling technologies that were not directly related to fiber waveguides, were also needed. The first was the development of small and relatively inexpensive semiconductor lasers, as light sources whose intensity can be turned up and down rapidly and efficiently. The first lasers had been made only in 1960, but these early devices were bulky and could only operate in a very inconvenient pulsed mode. They were therefore good for delivering large bursts of energy but were not much use for communication. The first semiconductor lasers appeared only in 1962 and it took at least another decade before the technology was sufficiently mature for lasers to be usefully connected to optical fibers. The second was the spread of computers from large organizations to the typical office and then the home, and now to individuals on the go, and with this, the shift of content from text to image to video, all of which caused the demand for information transmission to explode. Everything came together from the 1980s — much as envisioned in the 1966 paper.[8] Adoption went up and costs came down in a virtuous cycle. Kao wrote, in 1979,[15] just as the revolution was about to take off:

> "[T]echnological development is expected to proceed from strength to strength. However, the prognosis depends less on technical performance than on the cost reductions achievable. Costs decrease with production volume, yield and improved technology. All these [activities, applications, markets, and people involved] have been growing exponentially over the last few years […]."

Later technical improvements include optical fiber amplifiers which allow the signal to be boosted in strength as it propagates, and wavelength division multiplexing which allows many signals to be carried on the same fiber. Light carried through optical fibers is now literally the backbone of the Internet and of the information-rich world we live in today.

For his pioneering work that launched this revolution, Kao has received many awards, including a minor planet named in his honor, a British knighthood, and to cap it all off, the Nobel Prize for Physics in 2009.

Later Career

In parallel to the technical progress of fiber optics, the story of Kao's career should also be sketched. Four years after the seminal paper, Kao was recruited in 1970 to start the Department of Electronics[16] at The Chinese University of Hong Kong (CUHK). As the prospect of optical communication began to draw attention, ITT brought Kao back in 1974 to lead the division that would bring the technology to the market. There he stayed for 13 years, during a period when the field blossomed. In 1987, he was appointed Vice-Chancellor[17] at CUHK, which he led for nine years, through a period of expansion and advancement in research standards. The Faculty of Engineering was established during that time, with communication technologies being one of its core areas of strength. Those were politically tumultuous years in Hong Kong, but Kao remained the calm scholar with his eyes firmly set on long-term academic development. After his retirement in 1996, he took on various advisory roles in Hong Kong, especially in relation to the role of technology in the community. He was very much involved in the development of the Hong Kong Science Park, which now has an iconic lecture hall named in his honor. For a number of years, he lived in California, but he has now again returned to Hong Kong, making his home at CUHK.

Editors' Note

The editors are highly impressed by what Laureate Kao had to say from the following Baike website message: http://baike.baidu. <http://baike.baidu.com/view/2572889.htm.> com/view/2572889.htm.

> 事實上, 他亦從沒奪取過光纖技術的專利權, 光纖並沒有為他帶來巨大的財富 […]. "我沒有後悔, 也沒有怨言, 如果事事以金錢為重 […] 今天一定不會有光纖技術成果."

Translated, it reads:

> In fact, he has never garnered any fiber optic technology patents. His invention did not bring him any great wealth. […] "I have no regrets, no complaints. If every emphasis is about money […], there would not be any fiber optic technology achievements today."

October 10, 2018. The Editors mourn very much the untimely passing of Laureate Kao on September 23, 2018 in Hong Kong at the age of 84. It was almost 9 years ago on October 6, 2009 in the middle of the night when news from Stockholm of the Nobel Prize reached Professor Kao. Unfortunately, Professor Kao had been suffering from Alzheimer's Disease since 2004. Despite his illness, Professor Kao together with his wife created the Charles K. Kao Foundation to combat Alzheimer's Disease and raise public awareness about the disease. He is survived by his wife, Gwen, and their two children, Amanda and Simon. A public vigil on October 7 celebrated his life. Countless Hong Kong residents, officials, former colleagues and students came to pay final respects to Laureate Kao, the father of fiber-optics and the pride of Hong Kong. The memorial booklet distributed at the wake hailed Professor Kao as having "forever changed the course of human history." We are saddened that Professor Kao has taken leave from us and this world, but we will always remember his smile, his cheerful personality, and his momentous contributions. He will be greatly missed by all.

References

1. His given name in Chinese is "Kuen," a single character, and he also adopted the western name "Charles." Thus, in earlier years, "Kuen" came before "Charles," and he was known in publications as "KC Kao." As he spent more time in the West, he became generally known as "Charles," and his later papers were signed as "CK Kao" or "Charles K Kao." His passport puts "Charles" first, and he is of course now "Sir Charles," but his Social Security record shows "KC." This minor curiosity of the name symbolizes how he has moved across cultures and continents.
2. 'The Nobel Prize in Physics 2009'. Nobelprize.org. Nobel Media AB 2013. Web. 13 Nov 2013. http://www.nobelprize.org/nobel_prizes/physics/laureates/2009/

3. Kao CK, record of response provided to University of Greenwich Magazine for the purpose of an interview for a feature article (2007).
4. If multiple amplitude levels can be distinguished, then the information capacity is correspondingly increased. The ability to do so depends, in parts, on the signal-to-noise ratio.
5. The above is a simplistic account and the actual situation is worse than it may sound: (a) Various technical barriers would come in before hitting this theoretical limit, though these are not intrinsic and clever ways can be found around them. (b) To be more precise, the limit is set not by f itself, but by the range of f that is used, called the bandwidth. Thus, if the carrier wave occupies the range 1.00 MHz to 1.01 MHz, the limit is not a multiple of 1 M but a multiple of 0.01 M. (c) The capacity is expressed in the number of bits ("0" or "1") that can be transmitted per second, whereas nowadays we talk about bytes of 8 (or 16, even 32) bits. So 0.01 M bits per second translates to about 0.001 M bytes per second; a medium resolution picture of say 1 M bytes taken on a smartphone nowadays would then take ~10 minutes to transmit. (d) Worst yet, this is the overall capacity for everyone using this frequency, taken together.
6. A parallel investigation along this line, and in about the same period, bore unexpected fruit in another domain. Microwaves were employed for communication over medium distances, again using relays. There was a need to understand stray sources of microwaves that were getting picked up, because such "noise" sets a lower limit below which the signal could not be detected. In 1964, two scientists at Bell Laboratories, in trying to identify such background "noise," stumbled upon the cosmic microwave background which provided evidence for the big bang (now determined to be 13.75 billion years ago). Arnold Penzias and Robert Wilson received the Nobel Prize in Physics in 1978 for this discovery.
7. Kao CK, "Sand from centuries past: send future voices fast," Nobel Lecture (2009), in K Grandin (ed.) Les Prix Nobel. The Nobel Prizes 2009 (Nobel Foundation, Stockholm, 2010).
8. Kao KC, Hockham GA. (1966) Dielectric-fibre surface waveguides for optical frequencies. *Proc IEE* **113**(7): 1151–1158.
9. For the purpose of the present simple discussion, L is taken to be the distance for the power to drop to one-tenth.
10. Section 7 of Reference 8 refers to "the required loss figure of around 20 dB/km."
11. Section 3.1.2 of Reference 8: "The best absorption coefficient for glass is reported as … 200 dB/km."
12. Section 7 of Reference 8.
13. Section 4.4.1 of Reference 8.
14. Kao CK, Blanco CF and Asam A. (1984) Fiber cable technology. *J Lightwave Tech*, LT-2, 479–488. Table 1 therein cites a loss of 0.5 dB/km for single-mode operation at a wavelength of 1.3 μm.

15. Kao CK (1979) Optical fiber communication technology. *Electr Commun* **54**: 245–250; last section therein.
16. This later became the Department of Electronic Engineering, and one of the founding departments of the Faculty of Engineering when the latter was established in 1991.
17. For the benefit of those not familiar with British terminology, this post is equivalent to the President, with the Chancellor assuming a role that is largely nominal and ceremonial.

Other Recommended Readings

1. Kao CK. *A time and a tide: A memoir*. (The Chinese University Press, Hong Kong, 2011.)
2. Laureate Kao's autobiography: 高琨.《潮平岸闊 — 高錕自述》, translated from the above memoir by 許迪鏘 (Joint Publishing Co., Hong Kong, 2005. ISBN 978-962-04-2485-4).

Photo 9.1. Future laureate Charles K. Kao (middle) with his father, Chun-Hsiang Kao (right), mother, Jing Fang King (金静芳) (left); and younger brother, Timothy Wu Kao, 高鋙 (second right), taken in 1942. (Courtesy of Mrs. Gwen Kao.)

Photo 9.2. St. Joseph's College (聖若瑟書院) in Hong Kong, where Charles completed his secondary school education in the early 1950s. This prestigious school is run by the Institute of the Brothers of the Christian Schools (the Brothers also known as La Salle Brothers). (Courtesy of Dr. Hon-Lok Tang; photo taken in 2013.)

Photo 9.3. Class photo of Form 4B of Hong Kong's St. Joseph's College in 1950. Charles was at the first right of the third row. Mr. Quah (center, first row) was the class master. (Courtesy of Mrs. Gwen Kao.)

Photo 9.4. In 1966, at the Standard Telecommunications Laboratory in Harlow Town, England, young scientist Charles saw promise in a budding technology: optical fiber science. (Courtesy of CUHK.)

Photo 9.5. Charles receiving the Nobel Prize from King Carl XVI Gustaf of Sweden in 2009. (Photo credit: Jonas Ekströmer/SCANPIX/Sipa USA.)

Photo 9.6. The delivery of the Charles K. Kao's Nobel Lecture was assisted by Mrs. Gwen Kao (黃美芸), a talented engineer and an outstanding speaker, at Aula Magna in Stockholm University on December 8, 2009. (Copyright © The Nobel Foundation 2009, Photo: Orasisfoto.)

Photo 9.7. Charles and Gwen happily showing the Nobel Prize Medal and Diploma after the Nobel Prize Award Ceremony, 2009. (Courtesy of Mrs. Gwen Kao.)

Photo 9.8. A family portrait showing son Simon and daughter Amanda and taken at the time of the Nobel award ceremony. (Courtesy of Mrs. Gwen Kao.)

Photo 9.9. At the naming ceremony of Minor Planet (3463) "Kaokuen" by Nanjing's Purple Mountain Observatory in 1996. (Courtesy of CUHK.)

Photo 9.10. Shoulder to shoulder with students on the first day of university in 1994–95. (Courtesy of CUHK.)

Photo 9.11. Three Nobel Laureates at a lunch following the launch of the CUHK Nobel Exhibition. From Left to right: Professor CN Yang (Nobel Prize in Physics, 1957), Professor Charles Kao (Nobel Prize in Physics, 2009), Professor Sir James Mirrlees (Nobel Memorial Prize in Economic Sciences, 1996.) (Courtesy of CUHK.)

Photo 9.12. Ceremony for the Unveiling of the Statue of Sir Charles (Professor Charles K. Kao) in 2010 at The CUHK. From right to left: Professor Joseph Sung, Vice Chancellor and President, CUHK; Professor Wu Weishan, President of the Chinese Sculpture Academy, Chinese National Academy of Arts (Sculptor of the Statue of Professor Charles K. Kao); Professor Kao and Mrs. Gwen Kao. (Courtesy of CUHK.)

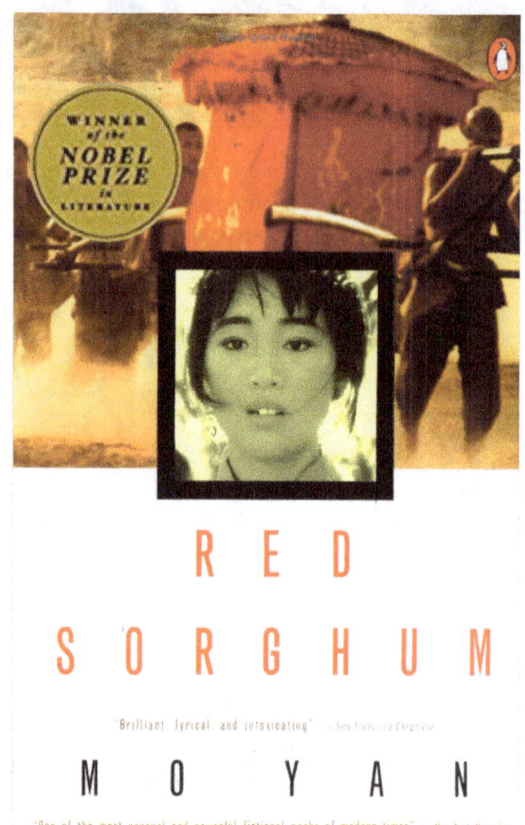

Book cover showing actress Gong Li (鞏俐). (Credit line: from *Red Sorghum: A Novel of China* by Mo Yan, translated by Howard Goldblatt, translation copyright © 1993 by Howard Goldblatt. Used by permission of Viking Books, an imprint of Penguin Publishing Group, a division of Penguin Random House LLC.)

Chapter 10

Mo Yan, 莫言
2012 Nobel Laureate in Literature

Victoria S. Lim

Mo Yan in Stockholm in 2012. (Courtesy of Laureate Mo Yan.)

The Nobel Prize in Literature 2012 was awarded to Mo Yan *"who with hallucinatory realism merges folk tales, history and the contemporary."*[1]

Since 1901 when the Nobel Prize in literature was first awarded annually, for a long 99 years, no Chinese had ever received the prize until Gao Xingjian (高行健) won in 2000. Unexpectedly, a short 12 years later, another Chinese writer, *Mo Yan*, again achieved that distinction. Such events are truly an unambiguous affirmation by the international community of the importance of Chinese literature to our global society.

Mo Yan (莫言) is a pen name; his legal name is "*Guan Moye* (管謨業)." His life and writing are tightly intertwined with and unequivocally shaped by the destiny of the Communist Party in China. He was born in 1955, a few years after the Communist Party won the Chinese Civil War and forced the *Chiang Kai-Shek's Kuomintang* government to exile in Taiwan.

His family, for three generations, was middle class peasants. They were not landowners and thus escaped being purged during the "Land Reform" and the "Anti-Rightist Movement" of the 1950s. Following the colossal failure of the "Great Leap Forward," China plunged into the "Great Famine" in the early 1960s when even the most fertile land was found untended. As a young child, *Mo Yan* experienced the unendurable distress of hunger. When his family had eaten all the leaves of the tree, they turned to eating the bark and then, finally the almost unchewable tough branches; one time they even ate coal pieces just to put some substances in their mouths. Clothing was equally sparse; *Mo Yan* and children of his age went naked for three seasons except during winter when they sought cover with almost unwearable rags.

When the Great Proletarian Cultural Revolution erupted in 1966, Mo Yan was 11-years old and was taken out of school because he was not among the poorest of the poor. He worked as a cowherd and later labored in factories. He recalled the secluded solitary existence in which he saw his lonely figure, stretched across the horizon, reflected in the dilated eye of the cow, his constant companion. He talked to the trees and the geese, but they did not respond and he finally learned to talk to himself. It was this childhood loneliness that led him to write his first successful novella, "*The Transparent Carrot.*"

In 1976 when the Cultural Revolution ended and Chairman Mao died, Mo Yan was 21-years old. He enlisted in the People's Liberation Army and began writing while he was a soldier. In 1986, he graduated from the People's Liberation Army Arts College and in 1991, obtained a master's degree in literature and art from the *Beijing* Normal University. In the late 1970s, after *Deng Xiaoping* started the Reform Movement, China entered a renaissance-like period during which *Mo Yan* was exposed to Chinese literature as well as the translated works of foreign authors including those of William Faulkner.

Faulkner had a particularly strong influence on him. *Mo Yan's* Northeast *Gaomi* Township (高密東北鄉) in Shandong Province parallels the Mississippi Yoknapatawpha County of Faulkner; each created his own fictional place.

In *The Transparent Carrot* the protagonist, a 12-year-old boy, fatherless and tortured by his alcoholic stepmother joins the great task of widening the reservoir for their township. He seems unable to focus on any work, cutting a finger while chiseling the stones and burning his hands while working with the blacksmiths. When his finger was bleeding, he sucks the blood with his mouth and covers the cut with mud. When his hands are burned, he runs down the bridge and immerses them in the river. Physically, he is stunted and malnourished, with every rib prominently protruding from his bare body. He stoically weathers the heavy chill of the late autumn wind. Socially, he is stifled, alienated and unable to form relationships. But he has a very active mind capable of vivid imagination. One night, after stealing a few carrots from the commune farm, there is one carrot left on the hot iron sheet. He sees shimmering blue and green lights emanating from the carrot, which has became transparent. That seems to make him happy, albeit for only a fleeting moment.

His subsequent novels became epic in nature. *Big Breasts and Wide Hips* covered China for the entire 20th century from the fall of the *Qing* dynasty to the end of the Cultural Revolution and then the Reform in the post-*Mao* era. It chronicles the long life of *Shangguan Lushi* from her birth to death. She is sold very young to marry a moronic and sterile husband and all her nine children — eight girls and one boy — are born through incest, and rape by Japanese soldiers, Chinese bandits and local peasants except the two youngest (twins) who are the product of a consensual affair with a Swedish missionary. She endures deep poverty and unbearable hunger and has to cope with the untimely and unnatural, not infrequently violent, deaths of her children and grandchildren. In the Post-Reformed era, while some of her descendants claim high government positions or became entrepreneurs, none of them end well because they are corrupt and greedy. At age 93, frail and destitute, she dies with her only surviving child — a son — who is poor and socially maladjusted, beside her. Readers would think such a pitiful life — a life embroiled in blood and filth — is not worth living, yet stoically, with immense resiliency and unbreakable strength, believing that Buddha decides her destiny, she lives stubbornly to nurture her grandchildren and her only son. Such is the power of the "breast," a symbol of "motherhood," and Mo Yan dedicated this book to the heavenly soul of his mother.

The Garlic Ballads is the story of peasants who planted garlic by the order of the government, but are, later on, unable to sell their crops despite taxes and bribe money when the storage house is full. Exhausted and hungry, they riot and the Party imprisons them. The prison is a place of filth and putrefaction; there are rats, maggots, lice and flies, and there are excrement and blood. The government is corrupt and brutal; peasants are of the sub-human species, identified by numbers and tortured in any way by the erratic discretion of the officials. Against this backdrop is intertwined a tragic love story. *Gao Ma* and *Jinju*, a farmer and a peasant's daughter, respectively, are, by any standard, a handsome couple, but their union is denied because *Jinju*'s father has sold her to an elderly sick farmer. They attempt to elope but are intercepted. Meanwhile *Gao Ma* is imprisoned after the garlic riot, *Jinju* suffers humiliation and unendurable poverty and hangs herself near the end of her pregnancy. After *Jinju*'s suicide, *Gao Ma* hears from a relative that their village has just performed a "wedding of the death" ceremony. A rich peasant's son has died, and his parents wanted their son to have a wife in the underground world. They pay *Jinju*'s brothers a large sum of money and dig out her remains, which are then transferred to the grave of the deceased son. *Gao Ma* is enraged, attempts an escape to reclaim *Jinju*'s remains, and is shot to death by the prison guards.

Life and Death are Wearing Me Out is a very creative piece of work. It tells the story of *Ximen Nao*, who at age 30 is shot by the Communist Party during the Land Reform. He complains to the underworld lord about his unjust treatment, and instead of going to heaven, he is reincarnated sequentially as five animals — a donkey, an ox, a pig, a dog, a monkey — and finally, a man again. Each animal is born to the same family of *Lan Lian* who was the hired hand of *Ximen Nao* during his life. *Ximen Nao* had three wives and *Lan Lian* wedded the second wife *Yingchun*. *Ximen Nao* left one son and a daughter — both conceived by *Yingchun* who subsequently had another son with *Lan Lian*. Thus the *Ximen* and the *Lan* clans are all blended together. Mystically, all the animals subconsciously know the history of the clan intimately. The major portion of the narration is done through the perspective of each of the re-incarnated creatures. They recount the story of the *Ximen* and *Lan* extended family spanning from the Land Reform to the year 2000 when a 4th generation infant was born on the eve of the new millennium. The boy, orphaned at birth — the father commits suicide and the mother dies of postpartum hemorrhage — is a product of parents who, unbeknown to them, share the same grandmother. As a result, the boy inherits a sex-linked blood clotting disorder; he will be taken care of by the grandparents. These creatures, thus, have witnessed the struggle and the humdrum of everyday living, the weariness of birth and death, the primordial lust of man and beast, the ugliness and flaws of human nature and the ever-complicated entanglement of human society. This story makes one wish for Nirvana, the state of enlightenment that breaks the otherwise endless rebirthing cycle.

The plot is complex and complicated; it deals with filial devotion, sibling rivalry, love and deceit, human kindness and cruelty, as well as government brutality and corruption. It deals with the transformation of the communist party's policy from anti-rightist, anti-intellectual, collective labor to one that stresses economic self-reliance and entrepreneurship. While life is unsparingly harsh in the beginning, it becomes luxurious at the end of the novel. Parked amidst the crowded shopping malls are Mercedes, BMWs and Cadillacs, women dressed fashionably, nails painted in vibrant red, some leisurely blowing out smokes from cigarettes. It is such a great metamorphosis.

Sandalwood Death is a historical fiction set during the Boxer Uprising when Chinese peasants resisted the occupation by the Germans. Under a complicated deal between *Qing* officials and the German general, an imprisoned peasant leader is slated for execution and this book describes the gruesome inhuman ways of killing with the singular purpose of inflicting the utmost pain. In this particular case, the executioner uses a cured sandalwood rod to skewer and pin the prisoner alive on a wooden stake and then force-feeds him strong medicine to keep him alive to prolong the suffering. Intertwined in this setting is the distasteful relationship of three families — that of the executioner, the prisoner and the county supervisor — and, in the end, death does not just occur in the prisoner, but also for the executioner and his son, and the wife of the county magistrate, all who perish violently.

Mo Yan had little formal education, but he educated himself well; flowing in his blood is 5,000 years of civilization from which he drew inspiration and energy. He is creative and imaginative and his writing explores many senses — sight, sound, smell and touch. I am particularly impressed by his description of all sorts of smell — body odor (living and dead), liquor and cigarettes, smog from car exhaust, unbearable stench from pigpens, offensive stink from filth and excrement and nauseating scent of blood from animals as well as humans. He always described scenery in meticulous detail — the sky, the sun and the moon omnipresent in any weather, the sorghum ubiquitous and the reeds abundant along the riverbanks.

In his heart, Mo Yan has a special venerable place for women; his female characters are resourceful and resilient, possessing immense capacity to withstand wear and tear. This feeling stemmed from his mother whom he spoke of with great adoration in his Nobel Lecture. I was most touched when he described his fear that his mother, weakened by hunger and disease and looking towards a future that is dark and bleak might commit suicide, and was much comforted when she tells him that there may be no joy in her life, but she would never leave him until the Underworld Lord calls for her. In his recent novel *Frog*, Mo Yan described with fond admiration his paternal aunt, a modern-day midwife, who was torn between her loyalty to the Communist Party's "One Child Policy" and her compassion for the peasant women whom she had to perform forced abortion at late-stage pregnancy, frequently with fatal outcomes. Such blind commitment to the ruling party not only blemishes her reputation, but also torments her with life-long remorse and prevents her from having a more fulfilled life. By contrast, many of his male characters are weak and loathsome. In *The Garlic Ballads*, Mo Yan described the despicable father and sons who treat *Jinju* like a piece of property, selling her body, alive and dead. *Jintong*, the son in *Big Breast and Wide Hips*, is painfully weak, docile and incapable. And the executioner in *Sandalwood Death* can methodically plan and torture a fellow human being without any single outward sign of agitation or fretfulness, like a cold-blooded animal.

Regarding the role of women in society, *Mo Yan* gives a contradictory picture. The mother in *Big Breasts and Wide Hips* is a traditional Confucian model. Despite the fact that her husband is sterile, she considers bearing her husband's family a male descendent her primary duty and has to steal the Y chromosomes from different men to bear eight daughters and, finally, the ninth child, a male infant. She is constantly subservient to the male members of the society and she endures abuse by her husband and torture by her mother-in-law impassively. On the other hand, the "Grandma" in the *Red Sorghum* is unconventional and very liberated. In defiance of the old Confucian teaching that a woman should not take charge but to obey her father, her husband and her son at different stages of her life, she strives to be the master of her own destiny. At age 16, sitting inside a suffocating sedan carried by four male bearers going towards the house of her husband-to-be who is totally unknown to her and a leper; her father had sold her for a head of big black mule. Her future looks grim and gloomy; she is alarmed and frightened and she decides there and then that she alone must take care of herself. For two nights, she refuses to sleep with her betrothed by carrying a pair of scissors. On the third day she rides a mule back to her parents' home for a traditional return visit. On the way, under the clear blue sky and beneath the shadows of the red sorghums, she offers her chastity to and consummates an adulterous relationship with a male sedan bearer — masculine and young — he is the "Granddad" in this story. Out of this encounter, she bears a son. While Grandma is away visiting her parents, Granddad secretly murders her legal husband and her father-in-law, giving Grandma the opportunity to inherit the sorghum winery, a tremendous financial asset that she manages admirably well. Granddad is a poor laborer, a bandit who later becomes her business partner in the winery and finally a hero fighting the Japanese army. Granddad feels no tinge of guilt and there is no adverse consequence to him from the murderous act. As for Grandma, has she sinned? I understand her fear of producing a "misshapen, putrid monster" and I do not blame her for using her own body as she thinks fitting. And who could fault her for wanting to live a more pleasant and happy life? In this story, the boundary between evil and goodness seems blurred.

What surprised me was the abundance of sex in his writing; sexual encounters — lawful or extra-marital — are so ubiquitous and primal. They occur inside the house, out in the yard, inside the office, along the riverbank, atop the tree branches and amidst the sorghum field. Mo Yan's language is raw and gripping and his detailed descriptions of sex, birth, illness and violent death are shockingly unflattering and make readers cringe. People in Mo Yan's work are full of character, some strong, others weak; all shamelessly exhibit primal human needs and desires and always flawed to some degree. He favors tragedy. Many characters in his novels die, some unavoidably, others, I think, unnecessarily.

Mo Yan's writing covers mostly the downtrodden people at the bottom of the society. He chronicles China during a century of miserable existence with war, deep poverty, great famine and a government that was brutal, corrupt and ignorant. His is an ugly and disgusting portrayal of the Communist Party, yet the authorities enthusiastically welcomed his Nobel Prize. This speaks well of the confidence of the current Chinese government. Gao Xingjian's books were banned in China, and *An Area of Darkness*, a book by V. S. Naipaul, another Nobel laureate, which negatively portrays his country during a return visit to his native land, was also banned in India.

Mo Yan's success is a result of the combination of inborn intelligence and hard work. He is not only creative, but also extremely prolific. As with any career, to be recognized requires an element of luck. Mo Yan is fortunate in that he had myriad, very capable foreign language translators such as Howard Goldblatt (English translations) and Anna Gustafsson Chen (陳安娜; Swedish translations). More importantly, he appeared on the scene at a strategic time when China was becoming an undeniable, rising power on the world stage, and the world was eager to learn more about her, including her contemporary writings.

Awards and Honors[2]

- 2005: Kiriyama Prize, Notable Books, *Big Breasts and Wide Hips*
- 2005: Doctor of Letters, Open University of Hong Kong
- 2006: Fukuoka Asian Culture Prize XVII
- 2009: Newman Prize for Chinese Literature, winner, *Life and Death Are Wearing Me Out*
- 2010: Honorary Fellow, Modern Language Association
- 2011: Mao Dun Literature Prize, winner, *Frog*

Editors' Note

Some people have voiced about Mo Yan's political inclination.[2] However, we are of the opinion that: (a) people are free to exercise choices, and (b) literature and politics are separate issues and should stay separate.

This is the first time that a Chinese citizen permanently residing in China has won the prestigious Nobel Prize in Literature for outstanding work carried out in China. Throughout its history, China has fostered scholarly pursuits and its citizens have derived great benefits from the stellar work of its literary giants. We look forward to the day when an increasing number of Chinese literature treasures are translated into myriad other languages for people of other countries to appreciate and enjoy.

The following list amply illustrates Mo Yan's exemplary literary talent[2].

Novels
- *Red Sorghum Clan* (紅高粱家族).
- *The Garlic Ballads* (天堂蒜苔之歌).
- *Thirteen Steps* (十三步).
- *The Herbivorous Family* (食草家族).
- *Big Breasts and Wide Hips* (豐乳肥臀).
- *The Republic of Wine: A Novel* (酒國).
- *Red Forest* (紅樹林).
- *Sandalwood Death* (檀香刑).
- *Life and Death Are Wearing Me Out* (生死疲勞).
- *Pow!* (四十一炮).
- *Frog* (蛙).

Short story and novella collections
- *White Dog and the Swing (30 short stories, 1981–1989)* (白狗秋千架).
- *Meeting the Masters (45 short stories, 1990–2005)* (與大師約會).
- *Joy* (歡樂) (8 novellas; six of them are published in English as *Explosions and Other Stories*).
- *The Woman with Flowers* (懷抱鮮花的女人) (8 novellas).
- *Shifu: You'll Do Anything for a Laugh* (師傅越來越幽默) (9 novellas; one of them, *Change*, is published independently in English).

Other works
- *The Wall Can Sing* (會唱歌的牆) (60 essays, 1981–2011).
- *Our Jing Ke* (我們的荊軻) (play).
- *Broken Philosophy* (碎語文學) (interviews, only available in Chinese).
- *Ears to Read* (用耳朵閱讀) (speeches, only available in Chinese).
- *Grand Ceremony* (盛典: 諾獎之行).

Books translated into English by Dr. Howard Goldblatt
- *Red Sorghum*.
- *The Garlic Ballads*.
- *Big Breasts and Wide Hips*.
- *Life and Death Are Wearing Me Out*.
- *The Republic of Wine: A Novel*.
- *Frog*.

Books translated into Swedish by Dr. Anna Gustafsson Chen

- *Red Sorghum* (Det röda fältet).
- *Garlic Ballads* (Vitlök Ballads).
- *Life and Death Are Wearing Me Out* (Liv och död bär mi gut).

References and Suggested Readings

1. Mo Yan — Facts — Nobel Prize. https://www.nobelprize.org/nobel_prizes.
2. Mo Yan — Wikipedia, the free encyclopedia. https://en.wikipedia.org/wiki/Mo_Yan Accessed on August 26, 2016.
3. Mo Yan — Biographical — Nobel Prize. https://www.nobelprize.org/nobel_prizes/literature/laureates.
4. Mo Yan — Nobel Lecture: Storytellers — Nobel Prize. https://www.nobelprize.org/nobel_prizes/literature/…/yan-lecture.html.
5. Mo Yan wins Nobel prize in literature 2012 | Books | The Guardian. www.theguardian.com.
6. Interview with Mo Yan (in Chinese) — Media Player at Nobelprize.org. www.nobelprize.org/mediaplayer/index.php?id=1889.
7. The Open University of Hong Kong Honorary Doctor of Letters Mo Yan wins 2012 Nobel Literature Prize. http://www.ouhk.edu.hk/wcsprd/Satellite?pagename=OUHK/tcGenericPage2010&c=C_ETPU&cid=191155253200&lang=eng.

Photo 10.1. Mo Yan's old residence in Gaomi (高密), Shandong Province, China. (Courtesy of Dr. Laurence K. Chan.)

Photo 10.2. Mo Yan, a soldier in 1976. (Courtesy of Laureate Mo Yan.)

Photo 10.3. Mo Yan and his mother in 1987. (Courtesy of Laureate Mo Yan.)

Photo 10.4. Mo Yan's father. (Courtesy of Laureate Mo Yan.)

Photo 10.5. Mo Yan receiving the Nobel Prize in Literature from King Carl XVI Gustaf of Sweden in Stockholm on December 10, 2012. (Photo credit: Jonas Ekströmer/SCANPIX/Sipa USA.)

Photo 10.6. Mo Yan poses after receiving his Nobel Prize from King Carl XVI Gustaf of Sweden during the Nobel Prize Ceremony at Concert Hall on December 10, 2012 in Stockholm, Sweden. (Photo credit: Pascal Le Segretain/Getty Images Entertainment/Getty Images.)

Photo 10.7. Left half of photo: Laureate Mo Yan and his wife Qinlan Du (杜勤蘭) with his Nobel Prize in hand in Stockholm on December 10, 2012. Right half of photo: Laureate Mo Yan was conferred the degree of Doctor of Letters, *honoris causa*, in 2005 by the Open University of Hong Kong [OUHK]. (Courtesy of the OUHK and of the Scanpix Sweden/Sipa USA.)

Photo 10.8. Mo Yan shows how to do Chinese caligraphy to Swedish students during his visit to the Hersby high school in Lidingo outside Stockholm, on December 7, 2012. (Photo credit: ANDERS WIKLUND/AFP/Getty Images.)

Photo 10.9. A woman places a poster of Mo Yan at the Unionsverlag booth at the 64th Frankfurt Book Fair on October 11, 2012. Mo Yan won the Nobel Literature Prize for his writings that mix folk tales, history and the contemporary, as announced by the Swedish Academy on October 11. (Photo credit: JOHANNES EISELE/AFP/Getty Images.)

Photo 10.10. Laureate Mo Yan at the International Writing Program of the University of Iowa in 2014. From left: Dr. Victoria Lim, Guan Xiao-Xiao (管笑笑; Laureate Mo Yan's daughter), Laureate Mo Yan and Dr. Ramon Lim. (Courtesy of Dr. Victoria Lim.)

Photo 10.11. Mo Yan and Dr. Howard Goldblatt at a seminar entitled "Chinese Literature as World Literature: Writer, Translator and Critics", held at the Open University of Hong Kong (OUHK) in Hong Kong in 2005. Dr. Goldblatt translated many of Laureate Mo Yan's books into English. Dr. Goldblatt was also honored with a Doctor of Letters, *honoris causa*, degree by the OUHK in 2008 for his outstanding literary accomplishments. (Courtesy of The Open University of Hong Kong.)

Photo 10.12. Dr. Anna Gustafsson Chen (陳安娜) translated several of Laureate Mo Yan's books into Swedish. (Courtesy of Dr. Anna Gustafsson Chen.)

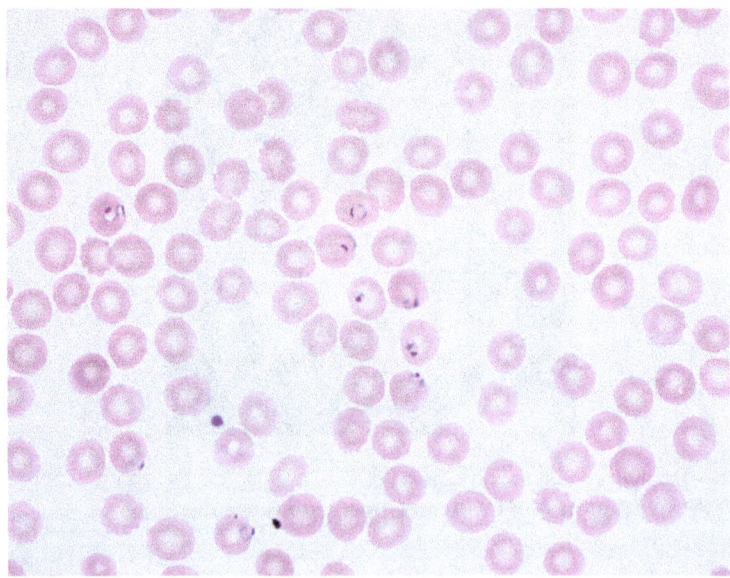
Red blood cells (clustered around the center of the photo) infected by the malaria parasite. (Courtesy of Dr. Hau C. Kwaan.)

Chapter 11

Youyou Tu, 屠呦呦
2011 Lasker~DeBakey Clinical Medical Research Award
2015 Nobel Prize in Physiology or Medicine

Ivan Barry Pless, Ray S. Huang, Xiao Qiang Ding and C. B. Lim

Laureate Tu receiving her Nobel Prize in Physiology or Medicine from King Carl XVI Gustaf of Sweden at the Stockholm Concert Hall on December 10, 2015. (Copyright © Nobel Media AB 2015. Photo: Pi Frisk.)

2011 Lasker~DeBakey Clinical Medical Research Award

From the Albert and Mary Lasker Foundation: Award given *"For the discovery of artemisinin, a drug therapy for malaria that has saved millions of lives across the globe, especially in the developing world."*[1]

2015 Nobel Prize in Physiology or Medicine

From the Official Web Site of the Nobel Foundation, Nobelprize.org. Prize motivation: *"For her discoveries concerning a novel therapy against malaria."*[2]

Early Life, Education and Enviornment

As noted well by the Chinese scientist, Chen Zhu, qinghaosu is the pride of China! [正如著名科學家陳竺所說: 青蒿素是中國的驕傲!] [from the back cover of Ref. 3]

Youyou Tu was born on December 30, 1930 in Ningbo (寧波), Zhejiang Province (浙江省), China.[3] She came from a scholarly and distinguished family. Her father was Tu Liangui (屠濂規) and her mother, Yao Zhongqian (姚仲千).

She attended the Xiashi Middle School (效實中學) in Ningbo where she met her husband, Li Tingzhao (李廷釗), a metallurgist, and they now have two daughters, Li Min (李敏) and Li Jun (李軍), one in Beijing and the other in the US. At an early age, Youyou became interested in both traditional Chinese medicine and western medicine. In 1951 she entered the Peking University School of Medicine (now known as the Peking University Health Science Center) to study in the Department of Pharmaceutical Sciences. She graduated in 1955, and later spent over two years in traditional Chinese medicine studies. After graduation she worked at the Academy of Chinese Medicine (now called the China Academy of Chinese Medical Sciences) and was promoted to a tenured researcher position in 1980. She is now the Chief Scientist at the Academy but it was only in 2011 after she had received the distinguished Lasker Award that Laureate Tu emerged from obscurity. Previously, her living conditions were spartan, as was her office, which was located in an old, poorly equipped apartment building in Beijing. According to Wikipedia,[4] her laboratory was prone to heating shortages and had only two electrical household appliances — a telephone and a refrigerator in which she stored her herb samples. In 2015, Youyou was awarded the prestigious Nobel Prize in Physiology or Medicine along with Satoshi Ōmura of Japan and William C. Campbell of the United States.[3-6] The latter two scientists discovered the chemical, avermectin (later modified as ivermectin), an agent effective against parasitic roundworms that cause lymphatic filariasis and river blindness.

Today, Laureate Tu is revered as a representative of the first generation Chinese medical workers since the establishment of the People's Republic of China in 1949 and, of course, as a leading international scientist.

Malaria, an Occasional Fatal Malady

Malaria is a disease that afflicts a large portion of the world's population, especially in certain hot geographic areas, and which can be fatal. The occasionally fatal and malignant form is caused by a parasite named *Plasmodium falciparum*. The vector is the female Anopheles mosquito. Because the disease was so deadly and widespread, it was the target of multiple efforts to eradicate it. For a while, with the help of the World Health Organization (WHO) and other organizations, it seemed that we were on the verge of conquering this disease just as we were with poliomyelitis. To the dismay of many people, however, the parasite gradually developed resistance to the most popular and previously effective drug, namely, chloroquine. Something new and potent was needed urgently to replace chloroquine. This is where Youyou came in and made her landmark discovery.[7-16]

Chairman Mao's Antimalarial Project 523

In 1967, at the instruction of the Chinese government under the direction of Chairman Mao Zedong, a research project called Project 523 — named after its starting date of May 23, 1967 — was launched to search for a drug to fight malaria in North Vietnam and in Southern China. Youyou's institute became involved also in trying to solve this problem and she was appointed to lead a talented team that included phytochemical and pharmacological scientists. The project altogether involved more than 500 researchers from about 60 institutions around the country.[13] What is special and unusual about Youyou's approach to this research challenge was the prominent role in which traditional Chinese medicine was destined to play.

In her Lasker acceptance speech Laureate Tu explained: "*In my childhood, I witnessed occasions when patients were rescued by folk Chinese medicine recipes. I started research on herbal medicines in 1955. My curiosity about herbs turned into a strong motivation after college training....*"[1] Thus, much of her training was at the Institute of Materia Medica where she spent two and a half years concentrating on traditional Chinese medicine. When she began applying this training to research in malaria, her team investigated more than 2,000 Chinese herb preparations and identified 640 that had possible antimalarial activities. Of these, about 380 extracts from 200 herbs were evaluated using mice infected with the malaria parasite.

A Classic Medical Text Leading to Qinghao and a Eureka Moment

Initially, there were no encouraging results. Then an extract of *Artemisia annua L.* (sweet wormwood; L. stands for Swedish scientist, Carl Linnaeus, a Father of Botany and the Father of Modern Taxonomy) showed some ability to inhibit the growth of parasites, but another setback followed. The initially encouraging observation could not be duplicated! Moreover, the initial findings seemed to contradict other reports. The team then carefully and thoroughly reviewed the literature. They found one reference relevant to the use of the qinghao (青蒿, the Chinese name of *Artemisia annua L.*) plant for alleviating malaria symptoms. The reference in the classic entitled *A Handbook of Prescriptions for Emergencies*, authored by Ge Hong (葛洪, 283–343 CE) stated that "A handful of qinghao immersed with 2 shēngs [升, one shēng is about one liter in volume[17]] of water, wring out the juice and drink it all".[18] This prompted Youyou to surmise that "the heating involved in the conventional extraction method that they had used might have destroyed the active components, and that extraction at a lower temperature might be necessary to preserve any antimalarial activity." This approach proved to be correct, and eventually, after further studies, WHO recommended the use of qinghaosu-based Artemisinin Combination Therapy (ACT) as the frontline remedy against malaria caused by *Plasmodium falciparum*. (Qinghaosu [青蒿素], the active ingredient of qinghao, has a chemical formula of $C_{15}H_{22}O_5$ and a molecular weight of 282 Da). During evaluation of the artemisinin compounds, it was found that dihydroartemisinin was more stable and 10 times more effective than artemisinin. More importantly, there was much less disease recurrence during treatment with this derivative. Adding a hydroxyl group to the molecule also introduced more opportunities for developing new artemisinin derivatives through the process of esterification.

Volunteering to be Human Guinea Pigs

After the successful extraction, Youyou herself volunteered to be the first experimental human subject. "As head of this research group, I had the responsibility," she said. It proved to be safe, so she then conducted human clinical trials that turned out to be successful. Her work was published anonymously in 1977. "It is scientists' responsibility to continue fighting for the healthcare of all humans," said Youyou. "What I have done is what I should have done in return for the education provided by my country."

Gift to Humankind

In her own words: "Equipped with a sound knowledge in both traditional Chinese medicine and modern pharmaceutical sciences, my team successfully accomplished the discovery and development of qinghaosu from qinghao." She concludes: "Whereas the finding of quinine was largely attributed to the historical use of *Cinchona ledgeriana* in Peru, the discovery of Qinghaosu is a gift to humankind from traditional Chinese medicine. Continuous exploration and development of traditional medicine will, without doubt, bring more medicines to the world. I advocate a global collaboration in the research of Chinese and other traditional medicines in order to maximize their benefits to the healthcare of human beings." Although she was grateful for the Lasker Award, she stated, "I feel more reward when I see so many patients cured." The awarding of the Nobel Prize, without a shadow of a doubt, attests to the enormous impact and importance of the discovery of Qinghao by Laureate Tu and her colleagues on the well-being of humankind.

Editors' Note

The discovery of artemisinin was an enormous, multifaceted and multi-centered endeavor, with the collaboration of a large number of scientists in China. Being the principal investigator of the project, Laureate Tu certainly deserves the lion's share of the credit.[15] Since the project was a multi-pronged effort, however, it is hoped that future recognitions will include some of the other scientists who have also contributed to the research undertaking.

This is the very first time that a stellar native Chinese scientist has earned not only the distinguished Lasker Award but also the prestigious Nobel Prize for research work carried out in China itself. It is our wish that the fruitful outcome of this challenging and momentous research project would serve to stimulate the development of more scientific research efforts in China and become a harbinger for future exemplary scientific breakthroughs emanating from 21st century China.

Myriad years of experience of the East when married to modern scientific know-hows of the West, along with back-breaking hard work, were instrumental in bringing about an unqualified and unprecedented accomplishment to Youyou's group. The latter was blessed by the best of both worlds! The success of Youyou's group encourages one to recall the trail-blazing achievements of Dr. K.K. Chen (陳克恢), a distinguished Chinese-American pharmacologist whose seminal work led to the use of ephedrine derived from a Chinese herb, "Ma Huang, (麻黃) (ephedra sinica)", in medicine in the first part of the 20th century.[19]

It is note-worthy that Laureate Tu and her associates employed the classic scientific method of analyzing the ingredients of a given substance and then sorting out those ingredients that were found to be promising, for further testing in the laboratory and in experimental subjects (including both animals and humans). Any approach that is less vigorous than the above regimen is unscientific and detrimental to the welfare of patients [possibility of death included].[20-22] To remain healthy, one is required and entitled to know exactly what one is taking into one's own body. Otherwise, one is just playing a dangerous game of Russian roulette!

Finally, we need to sympathize with the unfortunate victims of what has been called "Chinese herbs nephropathy or aristolochic acid nephropathy" due to the erroneous and unscientific intake of herbs containing aristolochic acid for the purpose of reducing weight or sundry other unscientific purposes.[23-25] Aristolochic acid nephropathy is characterized by inflammation of the interstitial tissue of the kidney, kidney fibrous tissue accumulation and atrophy, urinary tract cancers and eventual kidney failure.[23-25] Furthermore, aristolochic acid and its derivatives have recently been found to be a cause of liver cancer.[26] It is most heartening to learn that Laureate Tu and her group did the correct sequence of procedures in their qinghaosu studies and was on the right side of history all along. She and her outstanding colleagues wholeheartedly embraced the time-honored gold standard of performing medical research and held dear, at the same time, an exemplarily high caliber of medical care by observing meticulously the first principle of medical care, namely, "First, do no harm" — from Hippocrates, the Father of Medicine, 460–375 BCE.

Recently, a measure of resistance of the malarial parasite to the therapeutic effects of artemisinin has been detected in myriad countries. Combination therapies consisting of the simultaneous use of artemisinin and a variety of other antimalarial drugs have long been practiced to combat the emergence of such untoward resistance.[27-29] It is not inconceivable that complete resistance to artemisinin might raise its ugly head one of these days. Nevertheless, the most stellar research achievements garnered by Laureate Tu and her associates have been singularly instrumental in having saved millions of lives already. It is likely that this saving of lives will continue to take place for the foreseeable future, especially since various scientific measures are being intensively investigated with the aim of overcoming the resistance to artemisinin therapy. Even if complete resistance were to occur on a future date, the enormity of the beneficial influence on human lives as a result of this phenomenal discovery by Laureate Tu and her colleagues will go down in history as a highly celebrated event in the chronicles of humankind! From our point of view, Laureate Tu and her colleagues have earned their worldwide acclaim through their innovational talents and hard work. In our eyes, they are, without a doubt, internationally revered heroines and heroes for the ages!

Finally, the Editors found, from the Internet, some words of wisdom that are worthy of emphasis:

FROM: **CAIXIN** (A CHINESE ON-LINE MAGAZINE)
THE FICTION, THE FRICTION, AND THE FACTS FROM CHINA.
OCTOBER 23, 2015 1:27 PM.

All China Wanted Was a Nobel Prize[30]

by Zhang Yan

.........However, the Nobel Committee emphasized in explaining why Tu won the award that it was not given to traditional Chinese medicine but to a scientist who used sophisticated research methods to find a new therapy for malaria.

The process of turning an active component in plants into a modern drug involves laboratory studies and tests on animals, which is significantly different from how the herb was used in traditional Chinese medicine, Wang Liming, a professor at Zhejiang University's Life Sciences Institute, wrote in a commentary for *Caixin*.[30]

The following message pertaining to the awarding of the Nobel Prize to scientist Youyou Tu was obtained from **The New York Times**[31]:

"In fact, in its award, the Nobel committee specifically *said it was not honoring Chinese medicine*, even though Artemisia has been in continuous use for centuries to fight malaria and other fevers, and even though Dr. Tu said she figured out the extraction techniques by reading classical works. Instead, it said it was rewarding Dr. Tu for the specific scientific procedures she used to extract the active ingredient and create chemical drug."

Editors' Note

In common with the Nobel Committee[31] and Professor Wang Liming,[30] the editors believe that the Nobel Prize was meant to honor an exemplary pharmacognostical scholar who, along with her talented associates, successfully used a vigorous, modern, scientific approach to discover an antimalarial drug that can combat a malignant form of malaria, namely, the variety that is caused by *Plasmodium falciparum*. It cannot be overemphasized that the Prize was not awarded to traditional Chinese medicine. So any full embrace of traditional Chinese medicine by the world's scientific elites will have to wait for another day after outstanding scientific studies have given full credence to sundry herbal therapies.

Other Notable Awards[4]:

- 1978, National Science Congress Prize, P.R. China.
- 1979, National Inventor's Prize, P.R. China.
- 1992, One of the Ten Science and Technology Achievements in China, State Science Commission, P.R. China.
- 1997, One of the Ten Great Public Health Achievements in New China, P.R. China.
- 2011, GlaxoSmithKline Outstanding Achievement Award in Life Science.
- 2011, Outstanding Contribution Award, China Academy of Chinese Medical Sciences.

- 2012, One of the Ten National Outstanding Females, P.R. China.
- 2015, Warren Alpert Foundation Prize (co-recipient).

References

1. Strauss E. (2011) Award Description. 2011 Lasker-DeBakey Clinical Medical Research Award Recipient, Tu Youyou. *From the Lasker Foundation web-site, accessed on September 1, 2013.*
2. (a) Youyou Tu — Facts — Nobelprize.org https://www.nobelprize.org/nobel_prizes/medicine/laureates/2015/tu-facts.html

   (b) Youyou Tu — Biographical — Nobelprize.org https://www.nobelprize.org/nobel_prizes/medicine/laureates/2015/tu-bio.html
3. 屠呦呦傳：中國首獲諾貝爾獎的女科學家。屠呦呦傳編寫組。人民出版社。Youyou Tu's biography (TU YOUYOU ZHUAN).

   TuYouYou: The First Female Scientist of China Who Won the Nobel Prize, 2015. ISBN 978-7-01-015594-4.
4. *Tu Youyou, Wikipedia web site, accessed on April 25, 2017.*
5. Youyou Tu — Nobel Lecture. Nobelprize.org. Discovery of Artemisinin — A Gift from Traditional Chinese Medicine to the World.
6. The 2015 Nobel Prize in Physiology or Medicine — Presentation Speech. https://www.nobelprize.org/nobel_prizes/.../presentation-speech.html
7. Tu Y, Ni M, Zhong Y, *et al.* (1981) Studies on the constituents of *Artemisia annua* L. *Acta Pharmaceutica Sinica* **16**: 366–8 (in Chinese).
8. Tu Y, Ni M, Zhong Y and Li L. (1981) Studies on the constituents of Artemisia annua L. and derivatives of artemisinin. *China J Chin Mater Med* **6**: 31–2. (in Chinese).
9. Tu Y, Ni M, Zhong Y, *et al.* (1982) Studies on the constituents of *Artemisia annua L. J Medicinal Plant Res* **44**: 143–5.
10. Tu Y, Yin J, Ji L, *et al.* (1985). Studies on the constituents of *Artemisia annua L. Chin Trad Herb Drugs* **16**: 200–1. (in Chinese).
11. Tu, Y. (1987) Study on authentic species of Chinese herbal drug Qinghao. *Bull Chin Mater Med* **12**: 2–5. (in Chinese).
12. Tu Y. (2004) The development of the antimalarial drugs with new type of chemical structure — qinghaosu and dihydroqinghaosu. *Southeast Asian J Trop Med Public Health* **35**: 250–1.
13. Tu Y. (2011) The discovery of artemisinin (qinghaosu) and gifts from Chinese medicine. *Nat Med* **17**: 1217–20.
14. 屠呦呦。青蒿及青蒿素类药物 (QINGHAO JI QINGHAOSULEI YAOWU). 化学工业出版社, 北京, 2009. *Qinghao and Qinghaosu-similar Drugs* by Youyou Tu. ISBN 978-7-122-00857-2.

15. Miller LH, Su X. (2011) Artemisinin: discovery from the Chinese herbal garden. *Cell* **146**: 855–8.
16. Neill US. (2011) From branch to bedside: Youyou Tu is awarded the 2011 Lasker~DeBakey Clinical Medical Research Award for discovering artemisinin as a treatment for malaria. *J Clin Invest* **121**: 3768–73.
17. Volume of shēng (升). From Chinese Units of Measurement, Wikipedia website, accessed on October 18, 2013.
18. Ge Hong, Wikipedia website, accessed on October 18, 2013.
19. Chen KK, Schmidt CF. (1930) *Ephedrine and Related Substances* (Medicine Monographs XVII). The Williams & Wilkins Company, Baltimore.
20. Fan TP, Deal G, Koo HL, *et al.* (2012) Future development of global regulations of Chinese herbal products. *J Ethnopharmacol* **140**: 568–86.
21. Xu Q, Bauer R, Hendry BM, *et al.* (2013) The quest for modernisation of traditional Chinese medicine. *BMC Complement Altern Med* **13**: 132. Published online 2013 Jun 13. doi:10.1186/1472-6882-13-132.
22. Hurley D. (2006) *Natural Causes: Death, Lies, and Politics in America's Vitamin and Herbal Supplement Industry*. Broadway Books, New York, NY, 2006.
23. Pillans PI. (1995) Toxicity of herbal products. *NZ Med J* **108** (1012): 469–71.
24. van Ypersele de Strihou C, Vanherweghem JL. (1995) The tragic paradigm of Chinese herbs nephropathy. *Nephrol Dial Transplant* **10**: 157–60.
25. Yang HY, Chen PC, Wang JD. (2014) Chinese herbs containing aristolochic acid associated with renal failure and urothelial carcinoma: A review from epidemiologic observations to causal inference. *Biomed Res Int* **2014**: 569325. doi:10.1155/2014/569325. Epub 2014 Aug 27.
26. Ng AWT, Poon SL, Huang MN *et al.* (2017) Aristolochic acids and their derivatives are widely implicated in liver cancers in Taiwan and throughout Asia. *Sci Transl Med* **18**: 9(412). pii: eaan6446. doi: 10.1126/scitranslmed.aan6446.
27. Ashley EA, Dhorda M, Fairhurst RM, *et al.* (2014) Spread of artemisinin resistance in Plasmodium falciparum malaria. *N Engl J Med* **371**: 411–23.
28. Tun KM, Imwong M, Lwin KM, *et al.* (2015) Spread of artemisinin-resistant *Plasmodium falciparum* in Myanmar: A cross-sectional survey of the K13 molecular marker. *Lancet Infect Dis* **15**: 415–21.
29. Paloque L, Ramadani AP, Mercereau-Puijalon O, *et al.* (2016) Plasmodium falciparum: Multifaceted resistance to artemisinins. *Malar J* **15**: 149.
30. Liming W. In: Yan Z. All China wanted was A Nobel Prize. CAIXIN. The fiction, the friction and the facts from China. October 23, 2015. (CAIXIN is an Internet journal.)
31. Nobel Renews Debate on Chinese Medicine — The New York Times, https://www.nytimes.com/2015/10/11/…/nobel-renews-debate-on-chinese-medicine.html. Author: Ian Johnson, October 10, 2015.

Photo 11.1. Xiashi Middle School (效實中學) in Ningbo, the high school that Youyou attended. (Courtesy of Dr. Keith K. Lau.)

Photo 11.2. China Academy of Chinese Medical Sciences, where Youyou did her research work after her graduation. (Courtesy of Dr. Keith K. Lau.)

Photo 11.3. Youyou working in the laboratory after her graduation. (Courtesy of Laureate Tu.)

Photo 11.4. Youyou with mentor, Lou Zhicen (楼之岑), a pharmacognostical scientist, at the China Academy of Chinese Medical Sciences in 1951. (Courtesy of Wikimedia Commons.)

Photo 11.5. Laureate Tu (second left) with her husband Li Tingzhao (李廷釗) (right) and their first daughter Li Min (李敏) (second right) and second daughter Li Jun (李軍) (left). (Courtesy of Laureate Tu.)

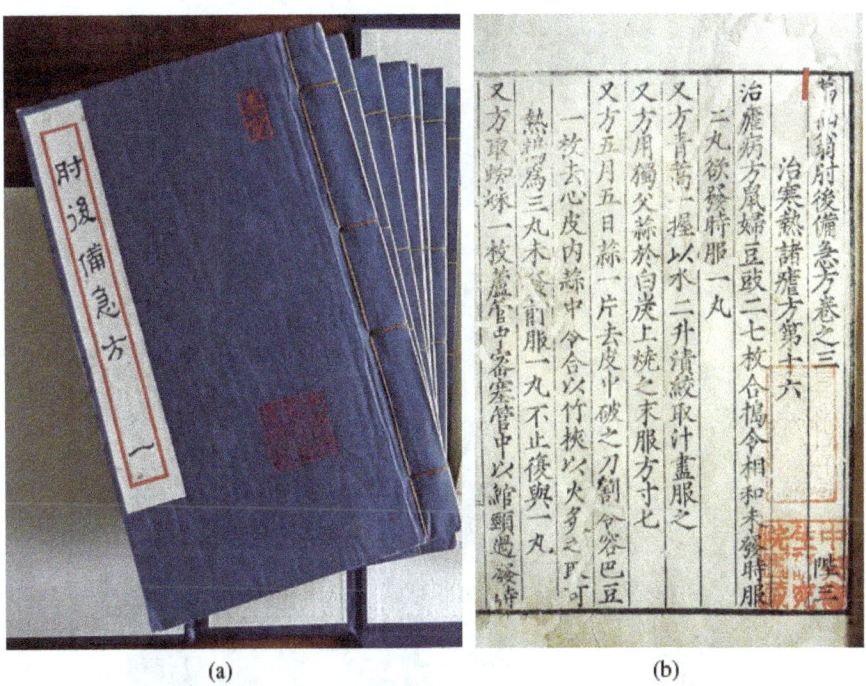

Photo 11.6. *A Handbook of Prescriptions for Emergencies* by Ge Hong. (a) Ming dynasty version (1574 CE) of the handbook. (b) The statement that "A handful of qinghao immersed with 2 "shēngs" of water, wring out the juice and drink it all" is documented in the fifth line from the right. (From Volume 3 of the Handbook). (From Ref. 13, used with permission.)

Photo 11.7. Ge Hong who composed "*A Handbook of Prescriptions for Emergencies*" during the Ming dynasty. (Courtesy of Mr. Alan Yan.)

Photo 11.8. *Artemisia annua L.* (a) A hand-colored drawing of qinghao in Bu Yi Lei Gong Pao Zhi Bian Lan (Ming Dynasty, 1591 CE). (b) *Artemisia annua L.* in the field. (From Ref. 13, used with permission.)

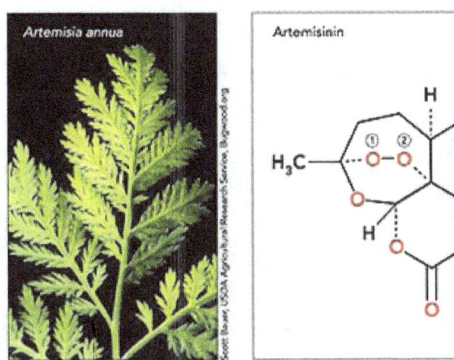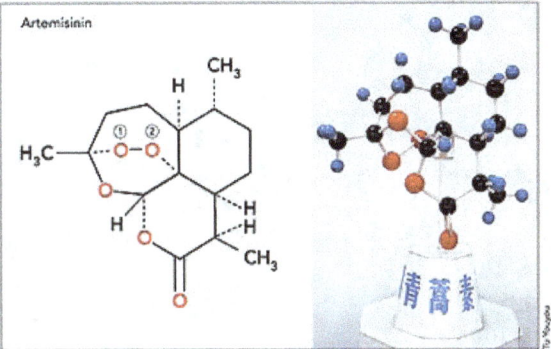

Photo 11.9. *Artemisia annua* (left), or sweet wormwood, contains the powerful antimalarial drug artemisinin, originally known as qinghaosu. The three-dimensional diagram of artemisinin (middle) shows the endoperoxide bond, which is crucial for the compound's antimalarial effects, between the numbered oxygens. The ball model of artemisinin (right) shows that bond on the left-hand side (two red oxygens linked to each other). The Chinese characters on the stand mean qinghaosu. (Courtesy of the Albert and Mary Lasker Foundation and of Ref. 13.)

Photo 11.10. Laureate Tu with her trophy after winning the Lasker Award (拉斯克獎) in New York City, USA, on September 23, 2011. (Courtesy of Laureate Tu.)

Photo 11.11. Laureate Tu at the 2011 Lasker Awards ceremony with the participants in a pre-ceremony discussion of careers in biochemical research. Sitting in the first row were: Arthur L. Horwich (left), John I. Gallin (second right) and Franz-Ulrich Hartl (right). Laureates Horwich and Hartl were fellow Lasker Award recipients. Dr. Gallin was the Director of the National Institutes of Health Clinical Center of the US. (Courtesy of the Albert and Mary Lasker Foundation.)

Photo 11.12. Youyou Tu at the Nobel Award Ceremony. (Copyright © Nobel Media AB 2015. Photo: Alexander Mahmoud.)

Lasker Laureates

"In quest for knowledge". (Photo courtesy of Mr. Fan Ho.)

Chapter 12

Choh Hao Li, 李卓皓
1962 Albert Lasker Basic Medical Research Award

Keith K. Lau, Cheryl D. Lau,
Daniel Tak Mao Chan and Sydney Chi Wai Tang

Laureate Li's portrait with his signature penned in 1956. (Courtesy of the National Library of Medicine, USA.)

1962 Albert Lasker Basic Medical Research Award Recipient. *For outstanding contributions to our understanding of the chemistry of pituitary hormones, including the identification and isolation of six hormones of the anterior pituitary gland (from the Albert and Mary Lasker Foundation, 1962.)*

Although no one can count how many Chinese have been recognized for their significant contributions to the advancement of sciences and medical knowledge, Professor Choh Hao Li is definitely among the most deserved. By the time he retired, he had made major contributions to our knowledge of a host of hormones from the anterior pituitary gland, the central nervous system and other tissues. These hormones include luteinizing hormone, corticotropin, growth hormone, follicle-stimulating hormone, prolactin, lipotropin, chorionic gonadotropin, insulin-like growth factor 1, melanocyte-stimulating hormone and β-endorphin. His pivotal works on growth hormone and β-endorphin have had a significant impact on myriad lives. In the modern era of clinical medicine, it is rare to find scientists who can surpass Choh Hao in contributions.

Choh Hao was awarded the Albert Lasker Basic Medical Research Award in 1962. His achievements ranged from contributions to the advancement of science, education and helping other academic centers (such as a center in Taiwan) to develop their own research institutes.

Early Life, Education and Family

Choh Hao was born in Guangzhou City (廣州市) of Guangdong Province (廣東省), on April 21, 1913. His father Kan-Chi Li (李鏡池) was a distinguished industrialist. Mew Shing Twui was his mother. Choh Hao finished high school at Puiying High School (培英中學) in Guangzhou in 1929. He then went on to study chemistry and obtained his B.S. degree in chemistry from the Jinling University (金陵大學) in 1933. Jinling University was established by the American Methodist Church in 1888, and was subsequently merged with the Nanjing University in 1952. He served as an instructor at the university for two years, then left China in 1935 for the USA to pursue his Ph.D. degree. During his tenure at Jinling University, he collaborated with Dr. W. V. Evans, a visiting Professor from Northwestern University, on a research study, and the results were published in the *Journal of the American Chemical Society*. He met his future wife, Shen Hwai Lu (Annie) (盧盛懷) in 1931 at Jinling University and they married in 1938 in the United States. Shen Hwai eventually obtained her master's degree in agricultural economics from the University of California at Berkeley. Subsequently, Annie and Choh Hao had a son, Wei-I (a thoracic surgeon) and two daughters, Ann-Si (a veterinarian) and Eva (an architect and environmental designer).

University of California at Berkeley

Choh Hao's long-term association with the University of California at Berkeley was not planned when he left China. His initial application to Berkeley was unsuccessful, probably due to the university's lack of knowledge about the academic standard of students from China at that time. While on his way to the University of Michigan to start his doctoral study, Choh Hao stopped by Berkeley in 1935 to visit his brother, Choh Ming Li (李卓敏). Choh Ming was a Ph.D. student in business administration at Berkeley at that time. [Choh Ming later became the founding Vice-Chancellor of the Chinese University of Hong Kong in 1963.] Professor Gilbert Lewis was the chief of Chemistry at Berkeley at that time. Choh Ming advised his brother to show his published paper (Ref. 9) to Professor Lewis (note: Li's name was spelled as Lee in this paper). Lewis was duly impressed by the paper's quality, and decided to accept Choh Hao into his program. By 1938, under the mentorship of Professor Thomas Dale Stewart, Choh Hao earned his Ph.D. degree in organic chemistry at UC Berkeley with studies focused on chemical kinetics. Choh Hao eventually joined the staff of Berkeley. He worked in the Institute of Experimental Biology under Professor Herbert M. Evans and rose through the ranks. During that time, Choh Hao embarked on the new field of human endocrinology, a largely unknown area of research. In 1949, he traveled to Europe to learn new techniques in biochemistry from Sanger and Porter in Cambridge (England) and Tiselius in Uppsala (Sweden). Upon his return to Berkeley, he was promoted to full professorship. He was appointed Director of the prestigious Hormone Research Laboratory (HRL) in 1954 and then dominated the field of hormone research for the next 40 years. The laboratory subsequently moved to the University of California at San Francisco in 1967 and Choh Hao continued to serve as the Director until his retirement in 1983.

Study of Hormones: An Under-explored Horizon

Due to a dearth of technology at that time, studying peptide hormones was not at all easy. When Choh Hao began his research endeavors on hormones, only a few laboratories could perform crystallization, centrifugation and moving-boundary electrophoresis. At that time, modern technologies such as amino acids sequencing, chromatography and gel electrophoresis had not been invented yet. Choh Hao, however, succeeded in meeting this hormone research challenge and proved to be most innovative.

His first significant research breakthrough was the purification and determination of the molecular structure of luteinizing hormone from the pituitary glands of sheep in 1940. A few years later, he succeeded in isolating another hormone, follicle-stimulating hormone and defining its interactive roles with luteinizing hormone in controlling the reproductive cycles of women. These findings were reported in the journal *Science* in 1949. In 1942, Choh Hao isolated corticotropin (also called adrenocorticotrophic hormone or ACTH) from sheep pituitary glands and published this seminal finding in *Science* as well. The amino acid composition of intact corticotropin was later identified and published in the *Journal of Biological Chemistry* in 1955. One common phenomenon that research scientists observed at that time was the overlapping biological activity of hormones, even in their pure forms. One typical example was corticotropin, which was also found to stimulate melanocytes. Unlike others, who attributed this phenomenon to cross contamination, Choh Hao hypothesized that the cross activity was likely due to the presence of homologous structures within different hormones. He also postulated that even part of an intact hormone was likely sufficient to elicit a measure of biological activity. This insight of Choh Hao's was proven correct when his laboratory discovered that the same 18 amino acid sequence present in melanocyte-stimulating hormone was also found within the 39-residue sequence of corticotropin. This outstanding and insightful speculation on the presence of active hormonal fragments inside a native intact hormone was confirmed repeatedly over the years.

Acquiring New Expertise from Europe

In 1957, Professor Li took a sabbatical to learn new peptide synthesis methods from Professor Robert Schwyzer in Basel. With the help of Professor Schwyzer, the Hormone Research Laboratory began to synthesize peptide hormones in 1958. Two years later, the team was able to synthesize the first 19 residues of corticotropin, which showed biological properties of both adrenal-stimulating and melanocyte-stimulating activities similar to those of the 39-residue of the native hormone. The entire human corticotropin hormone was synthesized successfully in the year 1973.

In 1975, Choh Hao's laboratory also discovered β-endorphins, isolating them from the pituitary glands of camels and confirming their analgesic effect. The laboratory continued to work on this hormone and eventually discovered the human counterpart of this hormone. Many of the non-addictive analgesics that are available today are the direct extension of his work on human β-endorphins. Besides his contributions to our current knowledge of pituitary hormones, Professor Li's research on growth hormone and insulin-like growth factor has also had a profound impact on how we manage many diseases today.

Choh Hao's legacy extends far beyond his research accomplishments. Since the establishment of the Hormone Research Laboratory in 1950 at Berkeley, the laboratory has been a much-sought-after institution for biochemical training. A host of successful scientists received their superior training from this world-renowned laboratory. For example, Professor Lin Ma (馬臨), the second Vice-Chancellor of the Chinese University of Hong Kong once spent nine months at this prestigious laboratory. Professor Ma still reminisces about those happy and productive days vividly even though it has been a lapse of more than 40 years already. He still remembers the times when Professor Li hosted dinner parties at his home, and how his wonderful wife, Annie, served delicious Chinese food. Choh Hao liked to collect antique china and enjoyed showing his treasured collections to his guests during those memorable occasions.

Signal Honors and Honorary Degrees

Among the numerous awards that Professor Li received, some are particularly worthy of mention; such as (a) the CIBA Award from the Endocrinology Society in 1947 acknowledging his promising research as a young scientist, and (b) the Koch Award by the same society in 1981 for his vast contributions to the field as a distinguished senior member. Professor Li received 10 honorary doctorate degrees from various academic institutions, including those from Hong Kong, Chile and Uppsala. He was also elected a fellow of many societies and academies, including the American Academy of Arts and Sciences, and the National Academy of Sciences.

Ingredients for Laureate Li's Momentous Achievements

Professor Li had an extremely successful career and people may wonder what contributed to his accomplishments. Professor Li's eldest daughter, Dr. Ann-si Li (李安熙), who has been practicing veterinary medicine in Hong Kong for the past several years, suggested that there were three main factors contributing to his success. Choh Hao had a very strong internal drive for success and always had the desire to achieve more. Another factor was the support of his wife who took care of everything at home so that he could concentrate on his work. In addition, she was always there to support him (*Editors' Note: Behind every successful man is a great woman*!). The third reason was his style of treating his fellows. Ann-si used the term "European-like" as the way Professor Li treated his fellows, similar to what one would take care of one's own family members. She could recall many social gatherings at their home. These gatherings were also the reason for her long-term friendship with a number of those fellows. Professor Lin Ma corroborated this during his interview with us. In the biographical memoir published by the National Academic Press, Professor D. R. Cole, one of his former fellows, commented: "My impression is that C. H. was successful in maintaining an ambience of joy, excitement, and strong friendship within the HRL throughout its history. The family-like affection of the large numbers who were associated with C. H. in the HRL was evident in enthusiastic reunions on the 20th and 30th anniversaries of the founding of the HRL, and on the celebration of C. H.'s sixtieth birthday." Finally, although Professor Li had an extremely successful career, he did not put pressure on his children, and left them a lot of freedom to choose their own career paths.

Although Professor Li passed away a number of years ago, his legacy survives. In 1987, a Chair in Biochemistry was endowed and a Lectureship was set up at the University of California at Berkeley. In Taiwan, there is an annual Choh Hao Li Memorial Lectureship, supported by a fund set up by Professor Li's eldest son, Dr. Wei-I Li (李偉怡), a thoracic surgeon at Bellevue, Washington, in association with the Academia Sinica, the Institute of Biological Chemistry, the Graduate Institute of Biochemical Sciences, the National Taiwan University and the Chinese Biochemical Society. This fund was established to honor Professor Li's leadership and his contributions toward promoting research in peptide hormone chemistry and biology in Taiwan. A special room housing a scientific library in his honor was also established at the Taiwan University.

Citation from the Chinese University of Hong Kong

Below is the Citation for Laureate Choh Hao Li (sometimes people put the surname Li first) on the Occasion of the 11th Congregation (1970) of the Chinese University of Hong Kong.

Dr. Li Choh-hao

Doctor of Law

Citation for Choh-Hao Li, Ph.D.

Dr. Li is a prominent chemist and experimental endocrinologist. He has spent thirty years conducting research on hormones, and is esteemed as the world's foremost authority on the subject.

Born of a distinguished family in Canton, Dr. Li first intended to study mathematics. Because of his liking for a certain professor, he changed his major field to chemistry. He earned the degree of Bachelor of Science from the University of Nanking in 1933 and was invited to teach chemistry at his alma mater after graduation. In 1935, he entered the University of California, Berkeley, and became its first graduate student from China to pursue advanced studies in chemistry. Three years later, he was awarded the degree of Doctor of Philosophy with flying colours. He also received an honorary degree of Doctor of Medicine from the Catholic University of Chile in 1962.

Dr. Li has a rich teaching and research experience. He has been research associate and assistant professor of experimental biology, associate professor and then professor of biochemistry and concurrently professor of experimental endocrinology at Berkeley. He has been both Founder and Director of the famous Hormone Research Laboratory, Berkeley and San Francisco, for the past two decades.

Numerous honours have been conferred on Dr. Li in recognition of his scientific achievements and contributions to mankind. He received the Ciba Award in Endocrinology in 1942, Guggenheim Fellowship in 1948, Francis Emory Septennial Prize of the American Academy of Arts and Sciences in 1955, Albert Lasker Award for Basic Medical Research in 1962, and Golden Plate Award of the American Academy of Achievement in 1964. He was also honoured by appointment as a Faculty Research Lecturer at the University of California, San Francisco, from 1962 to 1963, and as Annual Lecturer of the Japanese Endocrine Society in 1965.

Dr. Li has been elected to several prominent academic and professional institutions in the USA and other parts of the world, including fellowships in the American Academy of Arts and Sciences, the American Association for the Advancement of Science, and the Academy of Science. He has also been conferred honorary membership by the Argentina Society of Endocrinological Metabolism, Biology Society of Chile, and Academia Sinica of China.

Dr. Li has been an active member of the Advisory Board on Natural Sciences of the Chinese University of Hong Kong since 1964. In that capacity, he has greatly contributed to the development of our academic programme in chemistry. He is also the generous donor of the C. H. Li Funds for Research in Chemistry at this University. In deep appreciation of his services and contribution to our University, Li Choh Hao is being presented to Your Excellency as a candidate for the degree of Doctor of Laws, *honoris causa*.

(Above citation courtesy of the Chinese University of Hong Kong.)

Editors' Note

Needless to say that Laureate Li's achievements were not only exemplary but also phenomenal. Since the pituitary gland has been regarded as the Conductor of the Endocrine Orchestra, Laureate Li's momentous breakthroughs in the knowledge of the structures of the various anterior pituitary and other hormones have had immense influences on the better understanding of the physiological functions of those hormones, the manufacture of those hormones as well as the treatment of a host of disease states with those hormones. Among many others, an outstanding example of a clinical benefit is the use of growth hormone to foster growth in children suffering from growth hormone deficiency and consequent small stature. Another stellar example involves β-endorphin (the word endorphin was coined from endogenous and morphine), a neurotransmitter and a potent pain suppressor. It is only now, after so many years have elapsed since Laureate Li did his outstanding pioneering work, that we begin to appreciate the seminal significance of the hormonal actions of endorphins. We are now aware that β-endorphin is elevated in an array of important circumstances. The latter include pain, stress, exercise, acupuncture, opiate addiction and mental disorders.

References and Suggested Readings

1. CHOH Hao Li. *A History of UCSF*. University of California, San Francisco. http://history.library.ucsf.edu/li.html. Accessed on April 25, 2017.
2. Cole DR. (1996). *Choh Hao Li (1913–1987): A Biographical Memoir*. National Academy of Sciences. http://www.nasonline.org/publications/biographical-memoirs/memoir-pdfs/li-choh-h.pdf. Accessed on April 2, 2017.
3. Choh Hao Li, Biochemistry — California Digital Library texts.cdlib.org/view?docId=hb967nb5k3&doc.view=frames&chunk.id...
4. Hormone Research Institute. University of California, San Francisco. http://www.diabetes.ucsf.edu/research/research-units-resources/hormone-research-institute, accessed on November 26, 2013.
5. Li, Choh Hao. In: *The Asian American Encyclopedia*. Editor: Ng F. Publisher: Marshall Cavendish Corp, 1995, p. 988.
6. "Choh Hao Li", *Triangle; the Sandoz Journal of Medical Science*, **9**(1), pp. 41–2, 1969, ISSN 0041-2597, PMID 48969.
7. Forgotten Super Heroes of Science and Medicine: Choh Hao Li — Blogs. https://blogs.library.ucsf.edu/broughttolight/2016/09/15/choh-hao-li/.
8. Choh Hao Li. Wikipedia, accessed on April 25, 2017.
9. Evan WV, Lee FH, Lee CH. (1935) The decomposition voltage of grignard reagents in ether solution. *J Am Chem Soc* **57**: 489–90.
10. Li CH, Simpson ME, Evans HM. (1940) Interstitial cell stimulating hormone. II. Method of preparation and some physico-chemical studies. *Endocrinology* **27**: 803–8.
11. Li CH, Simpson ME, Evans HM. (1949) Isolation of pituitary follicle-stimulating hormone (FSH). *Science* **109**(2835): 445–6.
12. Li CH, Pedersen KO. (1952) Physicochemical characterization of pituitary follicle-stimulating hormone. *J Gen Physiol* **35**: 629–37.
13. Li CH, Simpson ME, Evans HM. (1942) Isolation of adrenocorticotropic hormone from sheep pituitaries. *Science* **96**(2498): 450.
14. Li CH, Geschwind II, Cole RD, Raacke ID, Harris JI, Dixon JS. (1955) Amino-acid sequence of alpha-corticotropin. *Nature* **176**(4484): 687–9.
15. Ramachandran J, Li CH. (1967) Structure-activity relationships of the adrenocorticotropins and melantropins: the synthetic approach. *Adv Enzymol Relat Areas Mol Biol* **29**: 391–477.

16. Farmer SW, Clarke WC, Papkoff H, Nishioka RS, Bern HA, Li CH. (1975) Studies on the purification and properties of teleost prolactin. *Life Sci* **16**: 149–58.
17. Li CH, Yamashiro D, Tseng LF, Loh HH. (1977) Synthesis and analgesic activity of human β-endorphin. *J Med Chem* **20**: 325–8.
18. Doneen BB, Bern HA, Li CH. (1977) Biological actions of human somatotropin and its derivatives on mouse mammary and teleost urinary bladder. *J Endocrinol* **73**: 377–83.
19. Gerner RH, Catlin DH, Gorelick DA, Hui KK, Li CH. (1980) β-Endorphin. Intravenous infusion causes behavioral change in psychiatric inpatients. *Arch Gen Psychiatry* **37**: 642–7.
20. Li CH, Yamashiro D, Gospodarowicz D, Kaplan SL, Van Vliet G. (1983) Total synthesis of insulin-like growth factor I (somatomedin C). *Proc Natl Acad Sci USA* **80**: 2216–20.
21. Li CH, Izdebski J, Chung D. (1989) Primary structure of fox pituitary growth hormone. *Int J Pept Protein Res* **33**: 70–2.
22. Choh Hao Li, Biochemist, Is Dead; Isolated Human Growth Hormone ... https://www.nytimes.com/.../obituaries/choh-hao-li-biochemist-is-dead-isolated-human-g...
23. The following is obtained from the distinguished Albert and Mary Lasker Foundation web-site.

1962 Albert Lasker Basic Medical Research Award

Isolation of six pituitary hormones

Choh H. Li
Hormone Research Laboratory University of California.

Photo 12.1. Laureate Li's favorite portrait (Courtesy of Dr. Wei-I Li.)

Photo 12.2. Puiying High School in Guangzhou. (Courtesy of Dr. Dekun Dong, 董德坤醫生.)

Photo 12.3. University of Nanking in 1920. (From: http://www.loc.gov/pictures/item/2007663089/ Library of Congress Prints and Photographs Division, Washington, D.C., USA.)

Photo 12.4. Laureate Li and Sir David Clive Crosbie Trench (Governor of Hong Kong and Chancellor of the Chinese University of Hong Kong) at the Chinese University of Hong Kong in 1970. (Courtesy of the Chinese University of Hong Kong.)

Photo 12.5. From right to left: Laureate Li, Sir David Clive Crosbie Trench and Professor Choh Ming Li (Laureate Li's elder brother and the Founding Vice Chancellor, the Chinese University of Hong Kong) at the Chinese University of Hong Kong in 1970. (Courtesy of the Chinese University of Hong Kong.)

Photo 12.6. From left to right: Laureate Li, Annie (Laureate Li's wife), Sylvia Zhi-Wen Lu (Professor Choh Ming Li's wife), Dr. Ann-si Li (Laureate Li's daughter) and Professor Choh Ming Li. (Courtesy of the Chinese University of Hong Kong.)

Photo 12.7. Recipients of the Honorary Doctor of Law, *honoris causa*, degree from the Chinese University of Hong Kong in 1970. From left to right: Mr. Ieoh Ming Pei (貝聿銘), Sir Sidney Samuel Gordon, Professor Choh Ming Li , Sir David Clive Crosbie Trench, Professor Tsung-Dao Lee (李政道), and Laureate Li. [N.B. Only Professor Choh Ming Li and Sir David were not recipients of the Law Degree.] (Courtesy of the Chinese University of Hong Kong.)

Photo 12.8. Photo taken in 1995 outside Dr. C. H. Li's Memorial Room in Taiwan. Besides his family members (wife Annie, son Wei-I, daughter Ann-si and granddaughter Lisa), many former fellows of Laureate Li's (including Professor Lin Ma, Professor Tung Bin Lo, Professor David Chung, Professor Laszlo Graf and Professor Janakiraman Ramachandran) were there in the photograph. (Courtesy of Dr. Wei-I Li.)

Photo 12.9. Laureate Li, explaining his favorite topic, namely, hormone chemistry. (Courtesy of the National Library of Medicine.)

Choh Hao Li (李卓皓): 1962 Albert Lasker Basic Medical Research Award Recipient

Photo 12.10. From left to right: Mary Lasker, Laureate Choh Hao Li (winner of the 1962 Albert Lasker Basic Medical Research Award), Robert Sarnoff, Laureate Joseph Smadel (winner of the 1962 Albert Lasker Clinical Medical Research Award) and Lady Bird Johnson (wife of President Lyndon B. Johnson). (Courtesy of the Lasker Foundation.)

The prestigious US National Institutes of Health Clinical Center in present days. Regarded as one of the best places to conduct biomedical research in the world. (Reprinted with permission from Macmillan Publishers Ltd: *Nature Medicine* **17**: 1221–3, copyright 2011.)

Chapter 13

Min Chiu Li, 李敏求 (1919–1980)
1972 Albert Lasker Clinical Medical Research Award

Steve Siu-Man Wong, Keith K. Lau,
Hung-Chun Chen, Carl M. Kjellstrand,
Antonios H. Tzamaloukas and Yuk-Lun Cheng

Dr. Min Chiu Li, a Father of Oncology.* (Photo reproduced with permission from the American Association for Cancer Research: DeVita V.T. Jr, Chu E. (2008) A history of cancer chemotherapy. *Cancer Research* **68:** 8643–53.¹)

*Oncology: In Greek, "onkos" means mass, bulk. Oncology means the study of tumors.

Albert Lasker Clinical Medical Research Award — Award Description of Min Chiu Li.[2]

For his outstanding contribution to the successful chemotherapeutic treatment of gestational choriocarcinoma.

Dr. Li (together with Dr. Roy Hertz), in 1956, was the first to demonstrate the first chemical cure of gestational choriocarcinoma resulting from pregnancy, by use of a drug — methotrexate.

Choriocarcinoma may take two forms — gestational, originating in the tissue that would normally be a constituent of the placenta, and non-gestational — i.e., that originating in the sexual glands of both sexes. Both types, when untreated, kill 90% of their victims.

Both types of tumors secrete a hormone that can be measured. Dr. Li and associates utilized techniques for measuring secretion of hormone as an index of tumor response to methotrexate. This enabled them to detect whether or not the cure was complete.

As a result, in 1961, Dr. Li and associates were able to initiate an improvement in the earlier treatment of gestational choriocarcinoma by using actinomycin D against those cases which had resisted treatment by methotrexate alone. The two drugs, later used sequentially by Dr. Li and other investigators, resulted in a 75% to 85%, 5 to 10 year cure rate, out of 500 patients so treated during the last decade.

The contributions of Dr. Li thus stand as an important landmark in cancer chemotherapy.

(N.B. Permission to reproduce the above description here has been granted by the Albert and Mary Lasker Foundation.)

Introduction

Dr. Min Chiu (Timothy) Li (1919–1980) was the first Chinese-American honored by the Albert Lasker Clinical Medical Research Award for his landmark achievements in cancer therapy. He was the first physician scientist to cure patients who suffer from metastatic gestational choriocarcinoma, using a chemotherapeutic drug called methotrexate.[3] Choriocarcinoma is a very rare placental cancer that afflicts pregnant women. The tumor was once considered universally fatal when it metastasized, and patients usually succumbed within six months.[4] Prompt surgical removal of the tumor was the only hope in the old days and most patients could only receive palliative care. Although a preliminary report had described the successful treatment of a patient suffering from metastatic choriocarcinoma with nitrogen mustard in 1954,[5] Min Chiu made history by demonstrating in more than one patient that disseminated choriocarcinomas could be cured definitively by drug treatment alone. Furthermore, he advocated, for choriocarcinoma treatment, the use of urinary excretion of a tumor marker, namely, a gonadotropin called choriogonadotropin[6–9] to monitor a patient's response to chemotherapy, thus ushering in a new important concept in modern cancer therapy that a tumor marker could be used to guide cancer management. As was the not uncommon fate of those who are way ahead of their times, Min Chiu's pioneering efforts were not well received by his peers in the oncology community. In fact, his novel and insightful approach to persevere with chemotherapy until normalization of urinary choriogonadotropin level was attained, in his patients with gestational choriocarcinoma who had initially responded to methotrexate therapy but had shown no discernible signs of tumor spread, was condemned as medical heresy that merited universal scorn. Unfortunately, instead of reaping well-deserved accolades, this innovative, forward-thinking and outstanding foresight eventually earned him an uncalled-for and unjust dismissal not only from the National Cancer Institute[1,4,8,10,11] but later also from the Sloan–Kettering Institute,[10] where he extended his work on testicular choriocarcinoma. Indeed it took many subsequent years for other investigators to affirm Min Chiu's ingenious, seminal therapeutic wisdom, i.e., in order to minimize the risk of recurrence, treatment needs to be continued and thorough until complete remission is at hand.[8] Undoubtedly, Min Chiu's contribution to modern oncology was unique and phenomenal. He brought the hope of survival to myriad "individuals with disseminated cancer." His form of treatment also allows the preservation of a patient's reproductive function by avoiding the surgical removal of treasured and irreplaceable reproductive organs.[3,12]

Growth of the Future Scientist

Born at a mission hospital on September 19, 1919, Min Chiu lived his early childhood with his mother, Mrs. Jeanette Li (Wen Wei-Chieh, 溫偉傑), in Tak Hing City (德慶市), a small city in Guangdong Province (廣東省) of Southern China. Life had never been easy for him. His father (Li Yung-Tsuen, 李榮焜) left the family to work far away in Hong Kong, leaving his wife and son "with no settled home, no anchor, no root."[13] Subsequently, there were also a few years when his mother was away, to complete her teacher training education, during which he was raised by his grandmother. When he reached his teens in the 1930s, Min Chiu left his ancestral home and moved with his mother to Nanjing (南京), where he attended high school. Meanwhile his mother, who was a devout Christian, moved northward to Manchuria in Northeast China in 1934 to fulfill her evangelistic calling as a teacher. He joined his mother in Manchuria in 1936.

Living in Mukden (now Shenyang [瀋陽市] of Liaoning Province [遼寧省]) of Manchuria, Min Chiu received his medical education at the Mukden Medical College. Founded by a Scottish missionary, Dr. Dugald Christie, in 1911, this college was remarkable not only for being the first Western medical school in Manchuria, but also in 1934 for earning the privilege of being the first foreign medical college to receive recognition by the University of Edinburgh.[14] Min Chiu stayed on in Mukden after graduation. At that time, a cholera outbreak had occurred and spread widely in Manchuria, and Min Chiu was commissioned to prepare antitoxins against cholera and other infectious diseases. As recalled by his mother, there was an incident when Min Chiu was very thirsty on a hot and busy day. He accidentally drank a cup of serum filled with cholera bacteria instead of clean water as he was preoccupied with his work. Thankfully he did not contract the disease, but that event attested to his passion for the sick and his commitment to his profession.[13] Indeed, during the Second World War, Min Chiu's working environment was never stable as he was working as a physician in a military camp under Japanese rule. His enthusiasm and care for the injured and the sick was highly appreciated. In particular, an imprisoned American soldier who had benefitted from his expert medical care, later returned his favors by collaborating with certain American religious benefactors, to help him to get to the United States to receive further medical training.[15] As a result, Min Chiu was able to move to the United States in 1947. Meanwhile, his wife, Pei Chia (百加), and mother stayed in China with his two children, a son and a daughter. Initially, Min Chiu studied bacteriology and immunology at the University of Southern California. Subsequently, he served as a medical resident at the Presbyterian Hospital (present day Rush University Medical Center) in Chicago in 1952.[3]

Development of a New Era in Chemotherapy for Malignancy

In 1953, Min Chiu joined Dr. Olof H. Pearson at the Sloan–Kettering Institute in New York as a Damon Runyon Fellow. His research involved the effects of chemotherapy on the reproductive organs, using a mouse model, especially utilizing placental tissues. One important insight that Min Chiu gained during this period was the potential linkage between methotrexate, a drug that inhibits the metabolism of folic acid, and choriogonadotropin production by placental tissues. His speculation was further strengthened by the dramatic reduction of urinary gonadotropin excretion upon an anecdotal methotrexate treatment given to a woman with metastatic melanoma who had her pituitary gland removed earlier.[6,7,16]

Methotrexate therapy in this patient was able to suppress the production of gonadotropin by the melanotic tissue (the latter now being the obvious source of gonadotropin production in the absence of a pituitary gland). Since both choriocarcinoma and testicular cancer can produce choriogonadotropin on account of the presence of chorionic tissues,[8,9] this correlation between urinary choriogonadotropin level and methotrexate

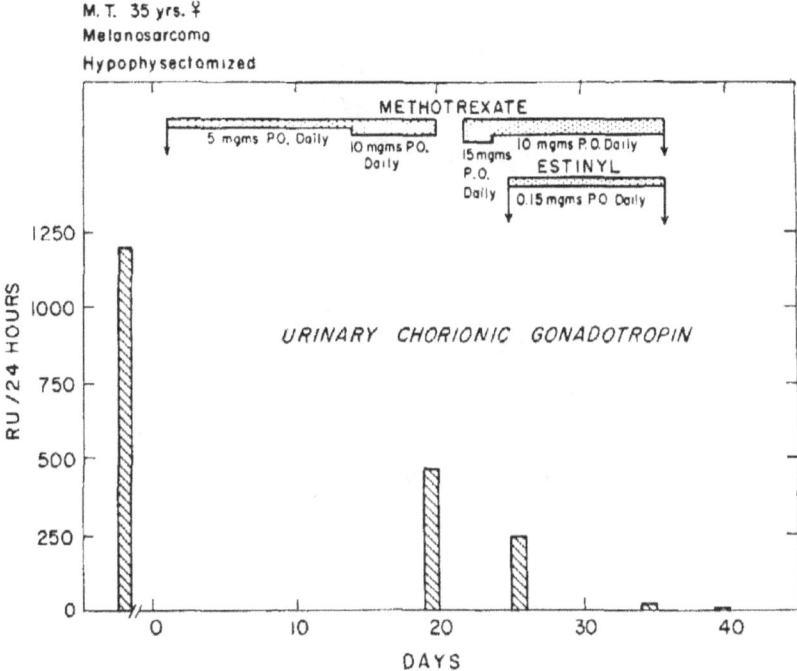

Fig. 1. The response of urinary chorionic gonadotropin in a woman with malignant melanoma treated with methotrexate. (Figure reproduced from "Li MC. in discussion, Hertz R, Bergenstal DM, Lipsett MB, Price EB, Hilbish TF. (1959) Chemotherapy of choriocarcinoma and related trophoblastic tumors in women. Ann N Y Acad Sci 80: 262–84",[6] with permission from Ann N Y Acad Sci. and John Wiley and Sons.)

therapy response eventually would become Min Chiu's cornerstone for the use of urinary choriogonadotropin as a tumor marker to monitor the effectiveness of treatment for these malignancies when using methotrexate or other cytotoxic drugs.[7,8]

In 1955, Min Chiu joined the National Cancer Institute (NCI), as a Clinical Associate in the Endocrinology Branch of the newly launched Clinical Center. This branch was headed by Dr. Roy Hertz at the time. Upset by the tragic death of a patient of his who had suffered from metastatic choriocarcinoma, Min Chiu proposed a research project utilizing methotrexate therapy for this devastating disease and gained approval from his superiors. In terms of using a cytotoxic agent to treat choriocarcinoma, it was known that, as reported by Anderson *et al.* earlier in 1954, a patient with choriocarcinoma and pulmonary metastases had responded to nitrogen mustard therapy.[5] To begin with, to seek expert advice for this historical project on the treatment of metastatic choriocarcinoma, Min Chiu consulted Dr. Emil J. Freireich, a highly experienced and well-respected investigator in leukemia therapy employing methotrexate and other drugs, and Dr. Paul Condit, who studied methotrexate pharmacology. Min Chiu, based on his own knowledge of the effect of methotrexate on the mouse reproductive organs, proposed a regimen of dividing a huge single dose into moderately-high doses

Emil Freireich

(Photo courtesy of the Albert and Mary Lasker Foundation.)

administered over several days. Indeed this approach marked the birth of the concept of intermittent intensive chemotherapy for cancer treatment that is still in vogue today. Min Chiu first started his methotrexate regimen in October 1955 on a 24-year-old woman with widespread choriocarcinoma. She developed a sudden rupture of her metastatic lung lesions that led to the filling of her chest cavity with blood and air (a condition known as hemopneumothorax). While the patient had been on the brink of death, upon methotrexate therapy initiation, surprisingly, she survived. Further intermittent doses of methotrexate were then given and the therapy resulted in a miraculous cure. It was notable that urinary choriogonadotropin excretion correlated well with clinical and radiological improvements during her methotrexate treatment. Following this unprecedented experience, Min Chiu treated some more similar patients and he was able to induce clinical remission by chemotherapy alone. These important findings were first published in the prestigious "Proceedings of the Society for Experimental Biology and Medicine" in 1956, with Min Chiu being the first author of the publication along with co-authors Hertz and Spencer:

> Li MC, Hertz R, Spencer DB. (1956) Effect of methotrexate therapy upon choriocarcinoma and chorioadenoma. *Proc Soc Exp Biol Med* **93**: 361–6.
>
> A part of the first page and a key figure (Fig. 2 below) of this article (Ref. 17) are reproduced below by permission from *Proc Soc Exp Biol Med* and *SAGE*.

Effect of Methotrexate Therapy upon Choriocarcinoma and Chorioadenoma.
(22757)

MIN CHIU LI, ROY HERTZ, AND DONALD B. SPENCER

National Institutes of Health, U. S. Public Health Service, Bethesda, Md.

Folic acid is known to be essential for the growth of the female genital tract and for normal embryonic development. Hertz et al. demonstrated the inhibition of estrogen-induced growth in the chick oviduct and monkey uterus in animals maintained on a folic acid deficient diet or treated with antagonists of either folic acid or adenine(1,2,3). Reduction of a previously high titre of gonadotropin occurred following a short course of methotrexate in a hypophysectomized woman with metastatic melanoma.*

The present report includes observations on the effect of a repeatedly administered and intensive regimen of methotrexate upon the urinary excretion of chorionic gonadotrophin and upon the clinical course in 2 patients with choriocarcinoma and one patient with chorioadenoma destruens.

Methods. The diagnosis was established by histological evidence. The initial extent of the disease was determined by physical examination and by radiological survey. Other observations included recording on each day of (a) fluid intake and output, (b) caloric intake, (c) the hemogram, and (d) the body weight. Each week x-rays were taken and measurement of palpable or visible lesions was recorded. Urinary gonadotropin determinations were carried out by a modification of the method of Klinefelter et al.(4). Gonadotropin values are expressed as "mouse uterine units" per 24 hours. The titres on

* Unpublished.

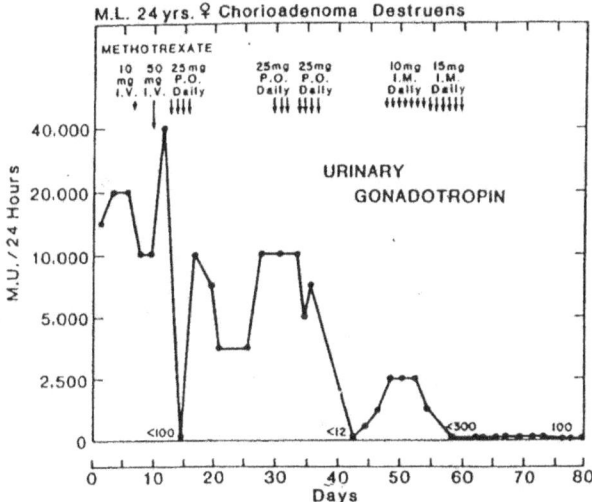

Fig. 2. Effect of methotreaxate on urinary gonadotropin excretion in a patient with metastatic uterine chorioadenoma. (From Li, M. C., Hertz, R., and Spencer, D. B.: Effect of methotreate therapy upon choriocarcinoma and chorioadenoma, *Proc. Soc. Exp. Biol. Med.* **93**: 361, 1956, published by Academic Press, Inc.)

[Favorable comments on the above article were made by: John Laszlo (1995) The cure of childhood leukemia into the age of miracles. Rutgers University Press, New Brunswick, New Jersey, p. 274, as follows: "Important progress in the treatment of a widespread cancer, which later would be the first to be cured by chemotherapy alone. Many lessons from the treatment of this rare cancer were applied to leukemia."[18]]

(*The above comments are reproduced with permission of Rutgers University Press, in the format Book via Copyright Clearance Center.*)

In addition, another landmark, more informative article on the same topic was published in 1958 by Min Chiu along with Hertz and Bergenstal in the influential *New England Journal of Medicine*[19]:

---

66   THE NEW ENGLAND JOURNAL OF MEDICINE   July 10, 1958

THERAPY OF CHORIOCARCINOMA AND RELATED TROPHOBLASTIC TUMORS WITH FOLIC ACID AND PURINE ANTAGONISTS*

Min Chiu Li, M.D.,† Roy Hertz, M.D.,‡ and Delbert M. Bergenstal, M.D.§

BETHESDA, MARYLAND

TUMORS of trophoblastic origin may arise either in the uterus or in the gonads. They are characteristically highly malignant and spread rapidly both by direct extension and by metastasis to lungs and brain and less commonly to numerous other sites.

We wish to report our experience with the treatment of 6 women and 5 men suffering from choriocarcinoma or related trophoblastic disease. We have previously reported our initial findings in 3 of these women and have also outlined our rationale for the use of an especially devised regimen of intensive, in-

completely subsided. Subsequent courses of five days' duration were repeatedly administered with similar precautions. The necessary interval between courses varied from seven to twelve days. Supportive medical and nursing management was provided throughout the treatment period. This included frequent oral lavage, parenteral infusion of fluids, extra nourishments and semiliquid diets when required.

The renal, hepatic and hematologic status of each patient was carefully assessed both before and at frequent intervals during therapy. Serial biopsies of

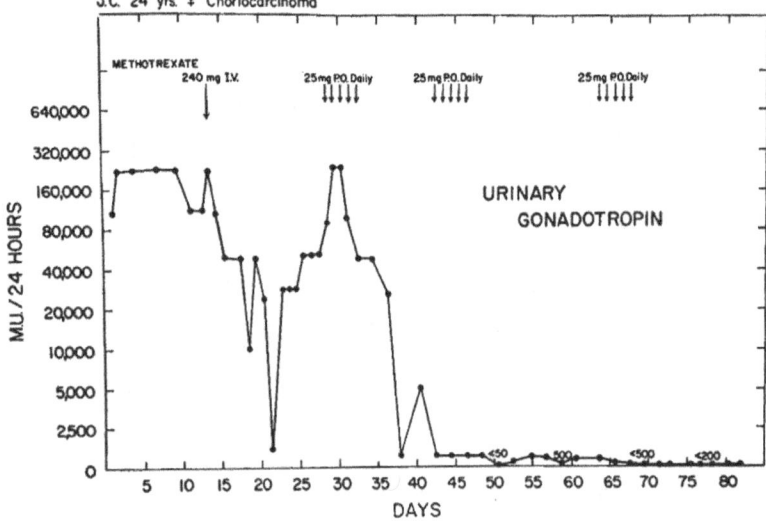

(From *The New England Journal of Medicine*; Min Chiu Li, Roy Hertz, Delbert M. Bergenstal; Therapy of choriocarcinoma and related trophoblastic tumors with folic acid and purine antagonists, Vol. 259, pp. 66–74. Copyright © (1958) Massachusetts Medical Society.)

(Above text and figure (from Reference 19) reproduced with permission from the Massachusetts Medical Society.)

The dramatic outcome of Min Chiu's work, as well as his unconventional approach to cancer therapy had understandably aroused intense interests and serious concerns within the NCI. Expert pathologists and clinicians, including some from Europe and Israel, were invited to review the available data and clinical management details from Min Chiu's patients. Queried by the experts on the possibility of spontaneous disease remission rather than genuine drug effect, as well as being challenged against the prolonged methotrexate exposure with concern about unnecessary drug toxicities, Min Chiu's findings were not accepted and a decision was made to repeat and validate his work on choriocarcinoma under alternate leadership. Sadly, he was asked to leave the NCI in 1957 before his innovative idea could be confirmed, appreciated and universally hailed by other investigators.

Nevertheless, Min Chiu continued to forge ahead with his research on cancer therapy in spite of the unfortunate setback. He returned to Sloan–Kettering Institute in New York in 1957 and kept exploring new treatment methods for both gestational choriocarcinoma in women and testicular choriocarcinoma in men. In sharp contrast to the women, response to methotrexate was much inferior in men and this observation led to his introduction of another important concept in modern oncology, namely, combination drug therapy. Realizing the heterogeneous nature of cellular tissues within a testicular choriocarcinoma, Min Chiu proposed in 1960 that various drugs with different mechanisms of action should be applied simultaneously to achieve a maximum therapeutic effect, while any overlapping of toxicities should be carefully avoided.[20] With such a strategy, much improved remission rates and patient survivals were had from clinical studies on testicular choriocarcinoma in the subsequent years. Moreover, a similar multi-drug approach was also proven useful when applied to gestational choriocarcinoma in women who had demonstrated resistance to methotrexate therapy alone. Since Min Chiu first demonstrated in 1961 that the combined use of actinomycin D and chlorambucil could bring about a complete remission in women suffering from choriocarcinoma who were resistant to single-agent chemotherapy,[12] combination chemotherapy for cancer treatment has become a common therapeutic tool in an oncologist's armamentarium these days. Apart from his work on choriocarcinoma, Min Chiu further published his study on colorectal cancer in 1976 when he was the Director of Medical Research at Nassau Hospital, Mineola, Long Island, New York. By showing improved survival with 5-fluorouracil in addition to complete surgical resection, he opened up another new avenue to optimize the outcome of colorectal cancer patients.[21] These pioneer studies formed the foundation on the roles of adjuvant chemotherapy that paved the way for other investigators in the field to extend the horizon of this important concept in oncology in the years to come.

Final Remarks

Without a doubt, Dr. Li is universally regarded as a pioneer of chemotherapy in modern oncology.[22,23] His exemplary work on choriocarcinoma treatment has allowed myriad patients with advanced disease to be cured with chemotherapy, with reduced risks of recurrence. Not only patients suffering from choriocarcinomas have survived with methotrexate therapy, but also some of those cured women have gone on to have healthy babies because the reproductive systems of these mothers have not been violated by either surgery or irradiation therapy. However, Min Chiu not only did not initially receive any credit for his ground-breaking work, a credit that he so richly deserved, but also, was wrongfully dismissed from his job for promoting precise and innovative therapeutic principles that were way ahead of his time. He got "discharged" for adhering to a trailblazing treatment approach that was eventually proved to be correct.[1,4,8,10,11] Nevertheless, Min Chiu accepted the adversity with great dignity and came out of it remarkably well. His incredible tolerance to setbacks and his ability to persevere in the face of adversities were not surprising, given the tough experiences that he had endured as a youth and the significant influence instilled into him by his inspirational mother.[13] Rather than being overwhelmed, hardship in Min Chiu's life immensely built him up and he emerged stronger, just like many others who successfully beat the odds.[24] The earning of the prestigious Albert Lasker Clinical Medical Research Award certainly brought forth a full vindication, attesting to the fact that Min Chiu's work was hugely appreciated and Min Chiu was on the right side of history all along.

In 1975, Min Chiu accepted the appointment as the Chairperson of the National Cancer Research Committee of the National Science Council in Taiwan.

In addition to the Albert Lasker Clinical Medical Research Award, Min Chiu was also honored with the New York Research Council Career Scientist Award in 1960.[25,26]

In conclusion, it is most heartening to learn that:

1. (a) In his description of how he figured out the way to treat children with leukemia successfully, Dr. Emil J. Freireich reminisced: "So we broke the prevailing concept of giving continuous low-dose treatment by treating the children aggressively even after all evidence of the leukemia was gone — I mean we treated with full doses. I would like to take credit for having this idea, but it really belongs to my colleague, Dr. M. C. Li. This man was a giant. He cured cancer; the first widespread cancer (choriocarcinoma) to be cured in the world was done by this man. Happily that fact was recognized his being given the Lasker Prize."[10] (*The cure of childhood leukemia: into the age of miracles by John Laszlo. Reproduced with permission of Rutgers University Press, in the format Book via Copyright Clearance Center.*)

    (b) In another interview, Freireich revealed his utmost admiration for Min Chiu: "But M. C. was even more radical than me. M. C. was the first person — you know it's amazing what he accomplished.... So he discovered the first tumor marker. He cured the first systemic human cancer. Fantastic breakthrough. That all went on right under my eyes. We talked to him every day, saw those ladies in the clinic. Of course Dr. Li was fired because he was too radical."[27]

2. While chronicling the initial use of chemotherapeutic agents at the NCI in its fight against cancer, Siddhartha Mukherjee had this to say in his masterpiece book[8]:

(a) In the case of choriocarcinoma (page 138):

"As Li had predicted, with several additional doses of methotrexate, the hormone level that he had so compulsively trailed did finally vanish to zero. While the patients who had stopped the drug early inevitably relapsed with cancer, the patients treated on Li's protocol remained free of the disease — even months after the methotrexate had been stopped. Not until several years later did it strike the board that had fired Li so hastily that the patients he had treated with the prolonged maintenance strategy would *never* relapse. This strategy — which cost Min Chiu his job — resulted in the first chemotherapeutic cure of cancer in adults."

(b) In the case of breast cancer (page 219):

"… a 33-year-old oncologist at the NCI, Paul Carbone, had launched a trial to see if chemotherapy might be effective when administered to women after an early-stage primary tumor had been completely removed surgically — i.e., women with no visible tumor remaining in the body. Carbone had been inspired by the patron saint of renegades at the NCI: Min Chiu Li, the researcher who had been expelled from the institute for treating women with placental tumors with methotrexate long after their tumors had visibly disappeared. Li had been packed off in ignominy, but the strategy that had undone him — using chemotherapy to 'cleanse' the body of residual tumor — had gained increasing respectability at the institute.…"

(*Reprinted with the permission of Scribner Publishing Group, a division of Simon & Schuster, Inc., and HarperCollins Publishers Ltd, from The Emperor of All Maladies: A Biography of Cancer by Siddhartha Mukherjee. Copyright © 2010 by Siddhartha Mukherjee, M.D. All rights reserved.*)

3. (a) Dr. V. T. DeVita, Jr, and Dr. E. Chu, at the National Institutes of Health, did pay Dr. Li an extraordinary compliment in the following message.[1]

"The very rare tumor of the placenta, choriocarcinoma, was the first to be cured. The preliminary results of a unique treatment program were reported in 1958. The principal architect of the treatment, using methotrexate in an unusual way for the time, was Min Chiu Li. The problem was no one was prepared to believe the results were significant because the primary site of the tumor was a parental hybrid tissue, subject, it was thought, to immunologic control. As a sign of the times, after the first two patients went into remission, they were presented at NCI Grand Rounds at the Clinical Center. The subject of the rounds was 'the spontaneous regression of cancer'. … Li was also told that if he persisted in using his radical treatment, he would have to forfeit his position at the newly opened clinical center. He persisted and was asked to leave"…

(*Reprinted with permission from the American Association for Cancer Research: DeVita VT Jr, Chu E. (2008) A history of cancer chemotherapy. Cancer Res* **68**: *8643–53*).

(b) In addition, on the occasion of the 17th research festival of the National Institutes of Health held on October 14, 2003, Dr. DeVita reminisced about the exemplary work performed by a deceased NCI colleague (Ref. 11; CC below means Clinical Center):

Daring Careers Remembered

Luminaries of CC Past and Present Launch 17th Research Festival

by Rich McManus

"Former NCI director (1980–1988) Dr. Vincent DeVita lingered for some time on the career of former colleague Dr. Min Chiu Li, who despite having discovered a cure for choriocarcinoma and other major research advances, was 'invited to leave' both NIH and another major medical center for the sin of being too far ahead of his time. DeVita seemed to suggest that such courage, married to such talent, is a rare thing these days, and ought to be cultivated."

Dr. Vincent T. DeVita, Jr. (Photo taken by Bill Branson; courtesy of Ref. 11.)

(c) Moreover, in an Oral History of his experience at the NCI,[28] Dr. DeVita reminisced:

..... He treated a dozen or so patients, and the results were the same — toxicity was fairly severe — and they fired him! They warned him that he just can't do this kind of outrageous thing, so they fired him. And he went to Memorial Hospital..........

The interesting thing is the first cure of testicular cancer was also done by M. C. Li, with a triple-drug combination — it was not a high cure rate, it was about a 20% complete remission rate, but those patients lived free of the disease. He was fired from Memorial Hospital for doing that, and then went into practice in Minneola, Long Island, where he just practiced Oncology and then retired in San Diego, and has since died. He did get recognition when he shared the Lasker Prize..........

4. On celebration of the 50th Anniversary of the Clinical Center's opening, Min Chiu's astonishing work was recalled by Pat McNees:

"*A young Chinese postdoctoral medical fellow, Min Chiu Li, brought from Sloan-Kettering some women with gestational choriocarcinoma, a rapidly fatal and rare cancer of fetal tissue of the placenta. Ann Plunkett, one of the first nurses on the cancer service, recalls, 'They would come in, these young women, and die within a matter of weeks to months.' Li proposed to Roy Hertz administering large doses of a new folic acid antagonist, known now as methotrexate, and was allowed to decide for himself whether to proceed. At first the drug made the patients ill; then one patient responded, and a second, and a third. 'It made you a real believer in medical research, to see these young women begin to live,' says Plunkett. In 1957, with single-agent chemotherapy, they had achieved not just remission, but a cure — the first successful chemotherapeutic cure for malignancy in a human solid tumor. Because it was an unusual tumor, with an immunological component (the placenta being considered the tissue the mother's body could reject), that first success was attributed to 'spontaneous remission.' Nobody would accept it as proof that chemotherapy could cure cancer, and Li was asked to leave NIH.*"[29]

5. The significance of Min Chiu's work was also highlighted in 2011 by Dr. John Gallin, the Director of the NIH Clinical Center:

"*In the 1950s, a team led by Min Chiu Li developed the first cure for a metastatic cancer (choriocarcinoma): a prolonged course of methotrexate. They were also the first to use a biomarker of cancer, human chorionic gonadotropin, to determine duration of treatment, and the method proved a more reliable predictor of treatment outcome than monitoring disappearance of symptoms. These monumental accomplishments, all by National Cancer Institute investigators working at the Clinical Center established that cancer can be cured, providing hope for patients and galvanizing the nation to increase support for cancer research.*"[30]

(Reprinted with permission from Macmillan Publishers Ltd: Nature Med *17*: 1221–3, copyright 2011.)

Editors' Note

The editors admire greatly Dr. Li's foresight, perseverance and never-give-up spirit; virtues essential for scientific success and succinctly emphasized by Nobel Laureate Steven Chu. Although Laureate Chu's advice conveyed below (please also see Chapter 5 of the present book) was primarily offered to physics-centric scientists, the general principles involved apply equally to scientists of other disciplines:

"On February 13, 2004 Steven gave an interview at the University of California, Berkeley, about his research and advice to scientists as the perquisites to do well in science. Steven stated that the physics scientist needs to have a strong mathematics background with a strong sense of natural curiosity and 'doggedness.' The scientist needs to have a strong internal drive and a passion to find out the answer. In Steven's experience, new discoveries are usually greeted with negative remarks from our colleagues who would say: It is wrong, it is trivial and you are not the first one to suggest this. However, one should not be discouraged and should continue to move on and find the truth".

The adverse experiences in research that Laureate Li went through is a shining testament to Laureate Chu's exquisite and inspirational advice. Although Laureates Li and Chu might not have met, they did share a very nice meeting of the mind.

Acknowledgment

We sincerely thank Dr. Emil J. Freireich, MD, Distinguished Teaching Professor at The University of Texas MD Anderson Cancer Center, Houston, Texas, for his invaluable information pertaining to this review.

References and Recommended Readings

1. DeVita VT Jr, Chu E. (2008) A history of cancer chemotherapy. *Cancer Res* **68**: 8643–53.
2. Albert Lasker Clinical Medical Research Award, Lasker Foundation. http://www.laskerfoundation.org/awards/1972_c_description.htm#li. Accessed on January 3, 2014.
3. Freireich EJ. (2002) Min Chiu Li: a perspective in cancer therapy. *Clin Cancer Res* **8**: 2764–5.
4. DeVita VT Jr, Goldin A. (1984) Therapeutic research in the National Cancer Institute. In: Stetten D Jr, Carrigan WT, editors. NIH: *An Account of Research in Its Laboratories and Clinics*. Academic Press, Inc., New York, NY, pp. 500–26.

5. Anderson HE, Bisgard JD, Greene AM. (1954) Metastatic chorionepithelioma of lung with nitrogen mustard therapy — preliminary report. *AMA Arch Surg* **68**: 829-37.
6. Li MC. In discussion, Hertz R, Bergenstal DM, Lipsett MB, Price EB, Hilbish TF. (1959) Chemotherapy of choriocarcinoma and related trophoblastic tumors in women. *Ann N Y Acad Sci* **80**: 262-84.
7. Li MC. (1979) The historical background of successful chemotherapy for advanced gestational trophoblastic tumors. *Am J Obstet Gynecol* **135**: 266-72.
8. Mukherjee S. (2010) *The Emperor of All Maladies — A Biography of Cancer*. Publishers: Scribner, New York, NY, pp. 135-8, 219.
9. Best CH, Taylor NB. (1945) The Physiological Basis of Medical Practice. 4th edition. The Williams & Wilkins Company, Baltimore, pp. 758-9.
10. Laszlo J. (1995) *The Cure of Childhood Leukemia: Into the Age of Miracles*. Chapter 8. Emil J. Freireich, M.D. Rutgers University Press, New Brunswick, New Jersey, pp. 108-55.
11. McManus R. (2003) Daring Careers Remembered. Luminaries of CC Past and Present Launch 17th Research Festival; NIH Record Vol. 55, No. 23 (November 11, 2003).
12. Li MC. (1961) Management of choriocarcinoma and related tumors of uterus and testis. *Med Clin North Am* **45**: 661-76.
13. Li J. (1971) *Jeanette Li: The Autobiography of a Chinese Christian*. Translated by Rose Huston. The Reformation Translation Fellowship. Publisher: The Banner of Truth Trust, London.
14. Crawford DS. (2006) Mukden Medical College (1911-1949): an outpost of Edinburgh medicine in northeast China. Part 1: 1882-1917; building the foundations and opening the college. *J R Coll Physicians Edinb* **36**: 73-9.
15. Xiao Min (1985) 同胞的愛 Tong Bao De Ai. Publisher: 九歌出版社 Jiu Ge Chu Ban She, Taipei, Taiwan, pp. 33-40.
16. Li MC. (1973) To the editor. *CA Cancer J Clin* **23**: 375-6.
17. Li MC, Hertz R, Spencer DB. (1956) Effect of methotrexate therapy upon choriocarcinoma and chorioadenoma. *Proc Soc Exp Biol Med* **93**: 361-6.
18. Laszlo J. (1995) *The Cure of Childhood Leukemia: Into the Age of Miracles*. Bibliography. Rutgers University Press, New Brunswick, New Jersey, p. 274.
19. Li MC, Hertz R, Bergenstal DM. (1958) Therapy of choriocarcinoma and related trophoblastic tumors with folic acid and purine antagonists. *N Engl J Med* **259**: 66-74.
20. Li MC, Whitmore WF Jr, Golbey R, Grabstald H. (1960) Effects of combined drug therapy on metastatic cancer of the testis. *JAMA* **174**: 1291-9.

21. Li MC, Ross ST. (1976) Chemoprophylaxis for patients with colorectal cancer. Prospective study with five-year follow-up. *JAMA* **235**: 2825–8.
22. Li, M. C. & G. B. Magill. (1958) Effect of folic acid and glutamine antagonists on chorionic gonadotropin producing tumors. Abstr Program of Endocrine Soc No. 53: 42.
23. Li MC. (1960) Current Status of Cancer Chemotherapy. *J Natl Med Assoc* **52**: 315–20.
24. Gladwell M. (2013) *David and Goliath*. Publishers: Little, Brown and Company, New York, NY, p. 158.
25. Min Chiu Li. Wikipedia, accessed on April 12, 2017.
26. Li JK. (2003) *CAMS at 40: 1963–2003, a History of Chinese American Medical Society*. Chinese American Medical Society, p. 20.
27. NCI Oral History Project Interview with Emil J. Freireich, M.D. http://history.nih.gov/archives/downloads/freireich.pdf. Assessed on January 3, 2014.
28. NCI Oral History Project Interview with Vincent T. DeVita, Jr., M.D. http://history.nih.gov/archives/downloads/devitainterview.pdf. Accessed on January 3, 2014.
29. McNees P. (2003) On the 50th Anniversary of the NIH Clinical Center's Opening. http://clinicalcenter.nih.gov/about/news/anniver50/_pdf/CC_50th_Anniversary_Celebration.pdf. Accessed on February 1, 2016.
30. Gallin JI. (2011) The NIH Clinical Center and the future of clinical research. *Nat Med* **17**: 1221–3.
31. Unknown authors (1992) 奉天医科大学 (辽宁医学院) 简史 Brief History of Mukden Medical College. Shenyang, Liaoning, China. (The library of the Chinese University of Hong Kong, Hong Kong, keeps a copy).
32. Li J. (2014) *Jeanette Li: A Girl Born Facing Outside*. An autobiography translated by Rose Huston. Second edition. Crown & Covenant Publications.

Photo 13.1. Min Chiu's life in China. (A) Tak Hing City (德慶市), where Min-Chiu was born in 1919. (B) Nanjing (南京), where Min Chiu attended high school. (C) Mukden (now Shenyang, 瀋陽市) where Min Chiu received his medical training.

Photo 13.2. Crest of Mukden Medical College (also spelt as Moukden Medical College), where Min Chiu received his medical training from 1940–1944. The crest consists of a needle, a snake and a torch: meaning: "To serve the sick and to support the injured with bright technology".[31] The four Chinese words at the center of the crest depict "To serve the people". The snake-entwined staff (staff of Asklepios [Latin, Aesculapius] is a popular symbol of medicine. The Chinese name of Mukden Medical College was changed to "盛京醫科大学" in 1933. (Photo courtesy of Mr. David Crawford, Emeritus Librarian, McGill University, Montréal, Québec, Canada).

Photo 13.3. The main building of the Mukden Medical College in the early 20th century. (Courtesy of Shengjing Hospital [盛京医院], Shenyang, Liaoning Province, China.)

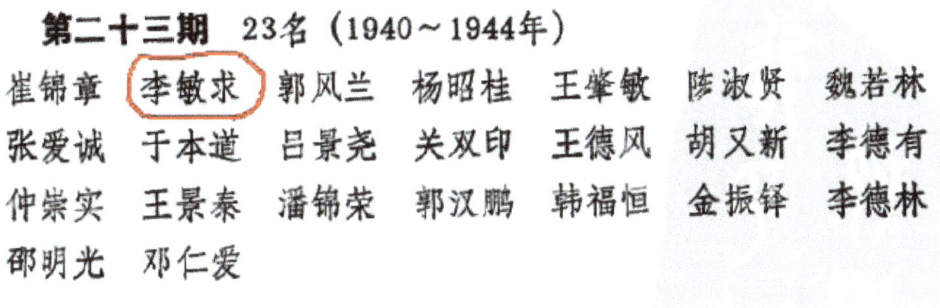

第二十三期　23名（1940～1944年）

崔锦章　李敏求　郭风兰　杨昭桂　王肇敏　陈淑贤　魏若林
张爱诚　于本道　吕景尧　关双印　王德风　胡又新　李德有
仲崇实　王景泰　潘锦荣　郭汉鹏　韩福恒　金振铎　李德林
邵明光　邓仁爱

Photo 13.4. Min Chiu (name encircled) and his 22 classmates in Mukden Medical College.[31]

Photo 13.5. Min Chiu in his younger days. (Courtesy of the National Cancer Institute, National Institutes of Health, USA.)

Min Chiu Li, 李敏求 (1919–1980): 1972 Albert Lasker Clinical Medical Research Award Recipient

Photo 13.6. Young Min Chiu with his mother, Jeanette Li. Photo taken in Qiqihar (齊齊哈爾), Heilongjiang Province (黑龍江省), China, in 1936–1937. (Photo from *Jeanette Li: A Girl Born Facing Outside* ©2014 Crown & Covenant Publications. Reproduced with permission.)[32]

Photo 13.7. An autobiography written by Jeanette Li, revealing her life as a devoted Christian, including stories about her son, Min Chiu. (Reproduced with permission from *Jeanette Li: The Autobiography of a Chinese Christian*. Translated by Rose Huston. The Reformation Translation Fellowship. Publisher: The Banner of Truth Trust, London. 1971.)[13]

Photo 13.8. Min Chiu with his son. (Reproduced with permission from *Jeanette Li: The Autobiography of a Chinese Christian*. Translated by Rose Huston. The Reformation Translation Fellowship. Publisher: The Banner of Truth Trust, London. 1971.)[13]

Photo 13.9. The Presbyterian Hospital in Chicago, where Min Chiu, along with Dr. Emil J. Freireich, received his residency training in the early 1950s. (Courtesy of the Facilities Photograph Collection, Rush University Medical Center Archives, Chicago, Illinois.)

Photo 13.10. The Sloan–Kettering Institute in New York, where Min Chiu worked from 1953–1955 and from 1957–1961. (Courtesy of Memorial Sloan Kettering Cancer Center.)

Photo 13.11. The old US National Institutes of Health Clinical Center where Min Chiu worked from 1955–1957. (Reprinted with permission from Macmillan Publishers Ltd: *Nature Med* **17**: 1221–3, copyright 2011.)

Photo 13.12. Lasker Award recipients in 1972 (with speakers and guests in brackets), from left to right: Back row: Emil Frei III, Vincent DeVita, Jr., V. Anomah Ngu, Donald Pinkel. Middle row: Min Chiu Li, Eugene J. Van Scott, Paul P. Carbone, John L. Ziegler, Joseph H. Burchenal, Denis Burkitt, Emil J. Freireich. Front row: Isaac Djerassi, Edmund Klein, (Mary Lasker, Sidney Farber, Alice Fordyce), C. Gordon Zubrod, Roy Hertz. An awardee, James F. Holland, was not in the photo. (Courtesy of the Albert and Mary Lasker Foundation; we wish to thank Laureates DeVita and Freireich for helping us to identify the distinguished individuals in the photo.)

Dr. Sun Yat-sen (孫中山), the Father of Modern China, is greatly revered and honored by the University of Hong Kong. Dr. Sun received his medical training from the Hong Kong College of Medicine for Chinese, which was amalgamated into the Faculty of Medicine of the University of Hong Kong in 1911. (Courtesy of the University of Hong Kong.)

Chapter 14

Yuet Wai Kan, 簡悅威
1991 Albert Lasker Clinical Medical Research Award

Susie Q. Lew, Hau C. Kwaan, Joseph M. Chan,
Angela T. Hadsell, Todd S. Ing
and Laurence K. Chan

Laureate Yuet Wai Kan receiving a Doctor of Science degree, *honoris causa*, from The Chinese University of Hong Kong on December 3, 1981. (Courtesy of The Chinese University of Hong Kong.)

Yuet Wai Kan was a young fellow in hematology at McGill University when he was asked to treat a premature newborn baby. Unfortunately, the baby ultimately succumbed to a condition called hydrops fetalis, but this tragic event spurred him to understand the root of the baby's demise. His subsequent research culminated in the characterization of sickle cell disease and thalassemia, which are inherited disorders of hemoglobin that constitute the most common genetic diseases in the world. These contributions had far-reaching consequences and led to the discovery of DNA polymorphisms, which formed the groundwork for the eventual advent of genetic testing and engineering. He was ultimately awarded the Albert Lasker Clinical Medical Research prize in 1991, but the events leading up to his accomplishments can be traced farther back in time even before his hematology fellowship, to his roots in Hong Kong.

"On a Slow Boat from China"[1]

Dr. Kan kindly invited one of the authors of this chapter to have lunch at Yank Sing, his favorite restaurant in San Francisco's Financial District. There, Kan shared many of his fond memories, particularly the guidance that he received from his former teachers and friends in Hong Kong.

Kan was born in Hong Kong in 1936, when the population of Hong Kong was just over half a million, a small fraction of the current seven million residents. This period was marked by the outbreak of the second Sino-Japanese War, with large numbers of refugees from mainland China flooding into Hong Kong to escape the conflict. Being the son of Mr. Tong-Po Kan (簡東浦), one of the founders of the Bank of East Asia, Kan was fortunate to receive a good local education. However, his studies at True Light Elementary School were often interrupted by intrusions during the Japanese occupation of the Second World War. To accommodate for these disruptions, his father would pay a private tutor for Kan's home schooling with bags of rice, as Hong Kong had no stable currency available during the war. After the war, he had his secondary education at Wah Yan College, a prestigious Jesuit high school known for producing notable alumni in the arts, sciences, business, and public service.

One of Kan's elder brothers, Sir Yuet Keung (簡悅強), was an eminent lawyer and legislator. Since Kan's older siblings had entered business, banking, insurance, and law, his father encouraged him to forge a different path and study medicine. In 1952, he was accepted to medical school at the Faculty of Medicine of the University of Hong Kong, the region's oldest institute of higher education. There, Kan joined the residence of Morrison Hall, where he immersed himself in an environment of unbridled creativity and academic enthusiasm among like-minded intellectual comrades. His clinical training took place mostly at the affiliate Queen Mary Hospital. After six years, Kan graduated with honors in 1958 with a Bachelor of Medicine, Bachelor of Surgery. After graduation, he completed his medical internship and residency at the Queen Mary Hospital under the direction of Professor A. J. S. McFadzean. It was an exciting time to be a house resident at Queen Mary Hospital and to learn from giants in the medical community like Drs. David Todd, Gerald Choa, Rosie Young and Stephen Chang. At the time, most young doctors were encouraged to perform research and study abroad for sub-specialty training, usually in the United Kingdom. Influenced in part by Professor McFadzean and his good friend David Todd, Kan took a different route and pursued the study of hematology in the United States.

In Boston, he joined Frank Gardner's laboratory at the Peter Bent Brigham Hospital where his research interests started to take shape. From Boston, he went to the University of Pittsburgh to complete his clinical training and next to the Massachusetts Institute of Technology (MIT) to study hemoglobin chemistry under Dr. Vernon Ingram. From MIT, Kan moved to Montreal where he obtained additional hematology training at the Royal Victoria Hospital of McGill University under Dr. Louis Lowenstein.

"Chance Favors only the Prepared Mind"[2]

Often in the history of medical discoveries, a chance but astute clinical observation can be the beginning of a fruitful, life-long scientific pursuit and discovery. Indeed, such was the case with Kan. During his first year of clinical work in Montreal in 1967, he saw a premature newborn patient who died from severe anemia.[3] He later found that this patient ultimately had a particular form of anemia called α-thalassemia. The diagnosis of this uncommon form of anemia at the time required an arduous process of protein analysis. As the pathogenesis of this disorder centers on the α-globin protein in red blood cells, Kan took the opportunity to study this protein when he joined his former colleague, Dr. David G. Nathan at the Boston Children's Hospital. By investigating the synthesis of this protein, they were able to identify many phenotypes of the disease.[4-9] At that time, molecular biology was in its infancy; most studies were carried out on non-primates and studies on humans were unheard of. Undeterred by this hurdle, Kan moved to the University of California at San Francisco (UCSF) and formed a team devoted to studying the structure of the α-globin gene. They discovered that the absence of the α-globin gene in the most severe form of α-thalassemia is lethal to fetuses before they are even born, leading to a fatal condition known as hydrops fetalis. They went on to pioneer the development of DNA techniques for the intrauterine diagnosis of this disease.[9-12] The same technique was also successful in the prenatal diagnosis of related disorders including β-thalassemia[13] and sickle cell disease.[14] These techniques are useful not only in the diagnosis of individual patients but also were employed in population-wide surveys. These studies showed that β-thalassemia (now also known as

Mediterranean anemia) is endemic to Mediterranean countries. Using DNA screening, thalassemic patients and carriers of the thalassemia gene were identified and were given genetic counseling. This groundbreaking approach is best exemplified by the successful reduction in the prevalence of thalassemia in Sardinia.[15-19]

Over the past five decades, Kan and his colleagues have published over 280 articles detailing their incredible findings. His earliest work on polycythemia in hepatocellular carcinoma, which inspired him to pursue hematology, was published in *Blood* (1961) with Drs. McFadzean and David Todd. His landmark study on the polymorphism of DNA was published in the *Proceedings of the National Academy of Sciences* (1978).[20] In the early seventies, when he was studying the DNA of a patient with both α-thalassemia and sickle cell disease using restriction endonucleases to cleave the DNA, he observed the presence of an abnormal fragment (a single nucleotide) in this patient's molecular pattern and went on to prove that this fragment is present in other patients with sickle cell disease.[20] This work, which is a technique widely used today, was one of the first studies to illustrate the linkage of single nucleotide polymorphism (SNP) to a disease. As of June 2015, approximately one and a half million SNPs are present in the *database SNP* from the National Center for Biotechnology Information (NCBI). Kan's discovery that a single nucleotide mutation can result in a change in the phenotype of a protein had a tremendous impact on the developing molecular science in the early seventies.

"Knowledge Comes, but Wisdom Lingers"[21]

On a more personal level, while Kan was at the Brigham, he met Alvera Limauro who was also working in Dr. Gardner's laboratory. After dating for two years, they married in 1964 in Boston and eventually had two daughters who would later bear them five grandchildren. After serving on the faculty at Harvard Medical School, Kan became Chief of the Hematology Service at San Francisco General Hospital in 1972 and a Howard Hughes Investigator in 1976. "That was a period of very rapid growth at UCSF," recalls Kan. At UCSF, he was part of a cadre of scientists that ushered in a new era of molecular biology and genetics that included future Nobel laureates Dr. Harold Varmus (now President of Sloan Kettering Institute and former director of the National Cancer Institute), Dr. J. Michael Bishop (former Chancellor of UCSF), and Dr. Herbert Boyer (cofounder of Genentech). Kan currently lives in San Francisco with Alvera, and he enjoys cultivating a small vineyard in St. Helena on weekends.

Kan received many national and international awards including the Albert Lasker Clinical Medical Research Award, the Gairdner Foundation International Award, and the Shaw Prize. He was also honored by the conferment of Fellowship status with the Royal College of Physicians of London and the Royal Society of London (1981 election), and Membership status with the United States National Academy of Sciences

(1986 election), the Academia Sinica (1988 election), the Third World Academy of Sciences (1988 election), the Institute of Medicine (US), and Chinese Academy of Sciences (foreign membership, 1996 election). He is the first Chinese elected to the Royal Society of London. The University of Cagliari in Italy, The Chinese University of Hong Kong, The University of Hong Kong, and The Open University of Hong Kong all presented him honorary degrees. He served as President of the American Society of Hematology in 1990, and he currently sits on both the Board of Adjudicators and the Selection Committee for Life Sciences and Medicine, which selects winners of the Shaw Prize. During the ceremony of the first Shaw Prize in Hong Kong, Kan publicly acknowledged the honor and his gratitude "as a native son of Hong Kong." He has indeed brought honor to the University of Hong Kong and the city of Hong Kong. For Kan, it has been an amazing journey and adventure.

"Ars Longa Vita Brevis"[22]

Kan's work in the molecular genetics of human blood disorders served as the foundation for much of the explosion of molecular diagnostics over the past 40 years, while his discovery of DNA polymorphism ultimately paved the way for the Human Genome Project. He continues his research (funded by Howard Hughes Institutes and by the National Institutes of Health) to find a cure for hemoglobinopathies using stem cells and gene therapy. His laboratory has successfully generated induced pluripotent stem (iPS) cells from mouse and human fibroblasts by retroviral delivery of transcription vectors.[23] His current work focuses on correcting mutations in these iPS cells and differentiating them into hematopoietic cells. A future treatment goal involves the removal of skin cells from a patient, differentiating them into iPS cells, correcting the mutations by homologous recombination, and differentiating them into the hematopoietic cells prior to re-infusion into the patient. Since the cells originated from the patient, there would not be any immuno-rejection.

Dr. Shinya Yamanaka, Professor of Anatomy at UCSF, and a recipient of the Shaw Prize in 2008 and the Nobel Prize in Physiology/Medicine in 2012, said of Kan: "I have high respect for Dr. Yuet Wai Kan and his science. Dr. Kan is one of the founders of modern genetics, and his contribution to our understanding of blood disorders has been enormous. More recently, he has applied iPS cell technology to his research. This research very much excites me, as it shows another beneficial application of iPS cells. As a person, Dr. Kan is polite, gentle, and brave. It is my pleasure and privilege to have worked with Dr. Kan."[24]

Kan has been a role model and a torchbearer for many young scientists at UCSF and worldwide. Another fellow prominent scientist, Dr. Dennis Lo, Li Ka Shing Professor of

Medicine at The Chinese University of Hong Kong, noted that early on as a medical student at Oxford, he was inspired by Kan's article, "On a slow boat from China," which documented his fascinating career in science. Despite his brilliance, Kan also embodies humility. On the nature of scientific discovery, he would observe, "People are just like cogs of the wheel. If the wheel loses a cog, it would still turn, although it might not turn as smoothly as before. And I am just like a cog of the wheel. This is what I think about my contribution." Others however see Kan's tremendous influence on this wheel of life as a spoke rather than a cog. As Dr. Stanley Prusiner, the 1997 Nobel Laureate for Physiology, said of Kan: "His work is the root, and from his work many, many people have elaborated on it."[25]

References

1. Kan YW. (1984) On a slow boat from China. Lita Annenberg Hazen award for excellence in clinical research. *Clin Res* **32**: 487–90.
2. Louis Pasteur: University of Lille Lecture 7, December 1854.
3. Kan YW, Allen A, Lowenstein L. (1967) Hydrops fetalis with alpha thalassemia. *N Engl J Med* **276**: 18–23.
4. Kan YW, Nathan DG. (1968) β-thalassemia trait: Detection at birth. *Science* **161**: 589–90.
5. Kan YW, Schwartz E, Nathan DG. (1969) Globin chain synthesis in the alpha thalassemia syndromes. *J Clin Invest* **47**: 2512–22.
6. Schwartz E, Kan YW, Nathan DG. (1969) Unbalanced globin chain synthesis in alpha-thalassemia heterozygotes. *Ann N Y Acad Sci* **165**: 288–94.
7. Kan YW, Golini F, Thach RE. (1970) A new protein synthesis factor from Escherichia coli. *Proc Natl Acad Sci U S A* **67**: 1137–1142.
8. Kan YW, Nathan DG. (1970) Mild thalassemia: The result of interactions of alpha and beta thalassemia genes. *J Clin Invest* **49**: 635–42.
9. Nathan DG, Lodish HF, Kan YW, Housman D. (1971) Beta thalassemia and translation of globin messenger RNA. *Proc Natl Acad Sci U S A* **68**: 2514–8.
10. Chang H, Hobbins JC, Cividalli G, *et al.* (1974) In utero diagnosis of: hemoglobinopathies. Hemoglobin synthesis in fetal red cells. *N Engl J Med* **290**: 1067–8.
11. Kan YW, Nathan DG, Cividalii G, Frigoletto F. (1974) Intrauterine diagnosis of thalassemia. *Ann N Y Acad Sci* **232**: 145–51.
12 Kan YW, Valenti C, Carnazza V, Guidotti R, Rieder RF. (1974) Fetal blood-sampling in utero. *Lancet* **1**: 79–80.
13. Kan YW, Golbus MS, Klein P, Dozy AM. (1975) Successful application of prenatal diagnosis in a pregnancy at risk for homozygous beta-thalassemia. *N Engl J Med* **292**: 1096–9.

14. Kan YW, Golbus MS, Trecartin R. (1976) Prenatal diagnosis of sickle-cell anemia. *N Engl J Med* **294**: 1039–40.
15. Kan YW, Lee KY, Furbetta M, Angius A, Cao A. (1980) Polymorphism of DNA sequence in the beta-globin gene region. Application to prenatal diagnosis of beta 0 thalassemia in Sardinia. *N Engl J Med* **302**: 185–8.
16. Trecartin RF, Liebhaber SA, Chang JC, et al. (1981) Beta zero thalassemia in Sardinia is caused by a nonsense mutation. *J Clin Invest* **68**: 1012–7.
17. Pirastu M, Lee KY, Dozy AM, et al. (1982) Alpha-thalassemia in two Mediterranean populations. *Blood* **60**: 509–12.
18. Pirastu M, Galanello R, Melis MA, et al. (1983) Delta +-thalassemia in Sardinia. *Blood* **62**: 341–5.
19. Cao A, Kan YW. (2013) The prevention of thalassemia. *Cold Spring Harb Perspect Med* **3**: a011775.
20. Kan YW, Dozy AM. (1978) Polymorphism of DNA sequence adjacent to human beta-globin structural gene: Relationship to sickle mutation. *Proc Natl Acad Sci U S A* **75**: 5631–5.
21. Lord Tennyson: Poems 1842.
22. Hippocrates: Aphorisms 400 BC.
23. Xie F, Gong K, Li K, Zhang M, Chang JC, Jiang S, Ye L, Wang J, Tan Y, Kan YW. (2016) Reversible immortalization enables seamless transdifferentiation of primary fibroblasts into other lineage cells. *Stem Cells Dev.* **25**(16): 1243–8.
24. Dr. Laurence K. Chan's personal communication with Dr. Shinya Yamanaka. February 23, 2015.
25. Statement made by Dr. Stanley Prusiner during an interview by Radio Television Hong Kong in 2005. 傑出華人系列. RTHK. (Message conveyed by Dr. Laurence K. Chan; photo courtesy of Radio Television Hong Kong.)

The Albert and Mary Lasker Foundation in 1991 gave the following reasons for presenting Yuet Wai "The Albert Lasker Clinical Medical Research Award":

"In pioneering studies using DNA to diagnose human disease, Dr. Kan has helped lay the groundwork for the rapid progress now being made in the identification of many genetic diseases and in prenatal genetic screening. His studies have markedly increased awareness of the potential use of recombinant DNA technology in diagnosis and have provided an extraordinary example of the power of molecular biology to bring about practical benefits. Dr. Kan's research has helped set the stage for the Human Genome Project.

In 1975, while analyzing the DNA of infants with α-thalassemia, Dr. Kan discovered that some were lacking the gene for the α-hemoglobin chain. This was the first demonstration of DNA deletion as a cause for human disease. Immediately, he applied this knowledge to developing a DNA test and in 1976, he successfully performed a prenatal diagnosis for α-thalassemia.

As recombinant DNA technology was being developed, Dr. Kan continued to devise applications for the diagnosis of human disease. He used restriction enzymes to cut DNA of patients with hemoglobin disorders into fragments for analysis. When comparing the fragment patterns from patients with and without sickle-cell anemia, Dr. Kan discovered that the patterns differed, thus providing a DNA test for sickle-cell anemia. This discovery of various patterns of DNA in 1978 has found broad application to DNA testing for many diseases. Today, scientists use techniques based on Dr. Kan's work using DNA samples taken from amniotic fluid to diagnose a range of genetic diseases before birth. These techniques are also used to locate genes on human chromosomes.

To Dr. Yuet Wai Kan, for his distinguished record of contributions to the understanding of hereditary diseases — in particular, his application of recombinant DNA technology for prenatal and predictive diagnosis — this 1991 Albert Lasker Clinical Medical Research Award is given."

http://www.laskerfoundation.org/awards/year/1991/

Other References

Kan YW — UCSF Profiles — University of California, San Francisco profiles.ucsf.edu/yuet.kan.

Yuet-Wai Kan — Autobiography — Life Science and Medicine — 2004. www.shawprize.org/en/shaw.php?tmp=3&twoid=53&threeid=68&fourid...

Yuet Wai Kan — Wikipedia. https://en.wikipedia.org/wiki/Yuet_Wai_Kan.

Photo 14.1. Yuet Wai received his elementary education from the "True Light School (香港真光中學及小學) of Hong Kong. Photo taken in 2016 shows the prestigious school in its new location at Tai Hang Road, Hong Kong. (Courtesy of Dr. Hon-Lok Tang.)

Photo 14.2. Yuet Wai attended Wah Yan College (華仁書院; center), a prestigious secondary school run by the Chinese Province of the Society of Jesus. To the lower right of the photo is the Catholic Cathedral of Hong Kong. (Courtesy of Laureate Kan.)

Photo 14.3. Queen Mary Hospital (瑪麗醫院), the main teaching hospital of the University of Hong Kong in the 1950's.

Photo 14.4. Yuet Wai (white arrow), along with his classmates, graduated from the Faculty of Medicine, The Univerisity of Hong Kong in 1958. (Courtesy of Drs. Eleanor K. Y. Ho and Paul W. K. Wong.)

Photo 14.5. A hematologist, a scientist, a scholar and the Chairperson of the Department of Medicine at the University of Hong Kong from 1948–1974, Professor A. J. S. McFadzean inspired Yuet Wai to follow his footsteps to become a hematologist, a scientist and a scholar himself also. Here at Ricci Hall of the University, Professor McFadzean succeeded in mesmerizing his adoring audience. (Courtesy of Dr. Bernard Tai.)

Photo 14.6. Posing for a casual, get-together photo in a hospital room. Back row: from left to right: Dr Yuet Wai Kan, a friend, Mrs. P. S. Kan, Dr. P. S. Kan and Dr. Kai Sum Lai. Front row: Professor David Todd. (Photo courtesy of Laureate Kan.)

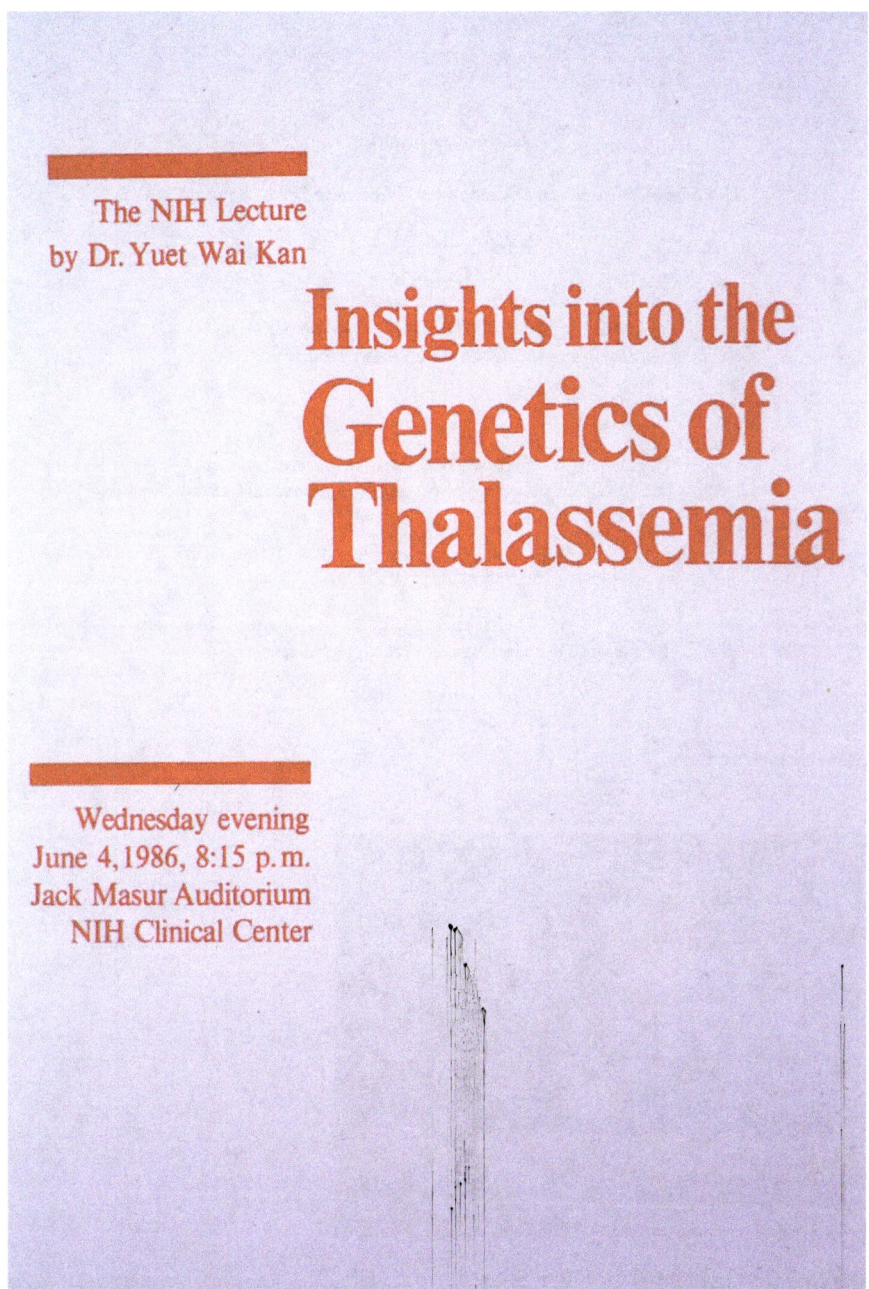

Photo 14.7. The US National Institutes of Health (NIH) lecture entitled "Insights into the Genetics of Thalassemia" was given by Yuet Wai on June 4, 1986. (Courtesy of US National Library of Medicine.)

Announcement

The Shaw Prize in Life Science and Medicine 2004 - Two Prizes

Prize One Laureates
Half jointly to

Stanley N Cohen & Herbert W Boyer
for their discoveries on DNA cloning and genetic engineering

Half to

Yuet-Wai Kan
for his discoveries on DNA polymorphism and its influence on human genetics.

Prize Two Laureate

Richard Doll
for his contribution to modern cancer epidemiology.

27 May 2004, Hong Kong

Photo 14.8. Announcement of The Shaw Prize in Life Science and Medicine for 2004 and Laureate Yuet Wai Kan accepting the Prize on May 27, 2004. (Courtesy of The Shaw Prize Foundation.)

Photo 14.9. Dr. and Mrs. Kan as guests of honor at the Hong Kong College of Physicians Annual Congregation and Dinner on October 13, 2012.
Front (L to R): R. Yu, Alvera Kan, Y. W. Kan, L. Chan
Back (L to R): R. Liang, J. Sung, P. Li, L. Yam, J. Tse, M. Ma. (Courtesy of Dr. Richard Yu.)

Photo 14.10. Yuet Wai Kan poses for a photo with his wife, Alvera, and Dennis Lo Yuk-Ming, FRS (recipient of China's 2016 Future Science Prize in Life Science) after attending the Symposium held at UCSF held on November 2, 2015 in honor of Dr. Kan's lifelong, momentous contributions in science and genetics. http://ihgsymposium.ucsf.edu/y-w-kan/.

Photo 14.11. Yuet Wai (with his signature at the bottom of the photo) was honored as a Chinese superstar by Radio Television Hong Kong. (Courtesy of RTHK,:香港電台.)

Photo 14.12. Photo of the September 4, 2016 Jubilee celebration of the birth of the University of Hong Kong Northern California Alumni Association. Yuet Wai (second row, R3) is revered everywhere by members of the University. Other acquaintances of the authors include Dr. Larry Ng (second row, L5), Professor Lap-Chee Tsui (second row, L6) and Dr. Kit Lau (President of the Northern California Alumni Association, second row, L7). (Photo courtesy of Drs. Larry Ng and Kit Lau.)

Special Feature, Wolf Laureate

Chien-Shiung along with some of her colleagues (2nd row, from left: E. Ambler, R. P. Hudson and D. D. Hoppes) from the US National Bureau of Standards at a press conference held on January 15, 1957, at the Physics Department located at Pupin Hall of Columbia University. The purpose of the conference was to announce the confirmation of the parity violation theory. To lend support, Professor Tsung-Dao Lee (front row) was also in attendance. [Courtesy of Tsung-Dao (T. D.) Lee Archive Online; copyright 2012 Shanghai Jiao Tong University.]

Chapter 15

Chien-Shiung Wu, 吳健雄 (1912–1997) 1978 Wolf Prize Laureate in Physics

Cheryl D. Lau, Kevin Chow, A. Ahsan Ejaz and Keith K. Lau

Laureate Chien-Shiung Wu (吳健雄). (Courtesy of the US Library of Congress).

Known as the "First Lady of Physics", whose most revolutionary finding contradicted the previously held popular notion that "weak" nuclear interactions between decaying particles are always symmetrical, Chien-Shiung Wu (Wu is the surname) has contributed significantly to the field of physics and is undeniably one of the most distinguished experimental physicists in modern history.

Wu was born on May 31, 1912 in the town of Liu Ho (瀏河鎮) of Taicang city (太倉市), in the Jiangsu Province (江蘇省) of China, which is near Shanghai. Wu's love of learning began with her parents, who filled their home with books, newspapers and magazines. Her parents often read with Wu, and encouraged her to be inquisitive and to think critically about the world around her. Her parents highly promoted education and were against gender inequality. Her father, Zhong Yi Wu (吳仲裔), was an engineer, a teacher and a school principal, and was involved in the 1911 revolution, which abrogated the Manchu rule in China. Wu's mother, Fuhua Fan (樊復華), was also a teacher. Together, her parents ran a girls' school called Mingde Women's Vocational Continuing School, which taught up to grade four. Wu attended this school until age nine in 1922, then enrolling in a boarding school named Suzhou Women's Normal School No. 2, which focused on teacher training based on a Western curriculum. There, Wu learned English from her classes, but also took the initiative to study science and mathematics independently, since these subjects were not offered in the curriculum. Besides studying, some of Wu's hobbies included writing and reading biographies. There was a law at the time that dictated that students who wish to pursue a university education must serve as a teacher for one year. Luckily for Wu, this law was not enforced during her year; so Wu instead took some courses at the National China College because Suzhou did not offer enough courses to satisfy Wu's intellectual curiosity. Wu attended a lecture by Shih Hu (胡適), a famous scholar and writer in China who had a profound influence on her. Hu was so impressed with Wu's intelligence and deep understanding of difficult philosophical concepts that he kept in touch with her beyond her graduation from the school. Wu admitted that other than her father, Hu had the most influence on her life. She graduated from Suzhou as valedictorian in 1930.

After high school, Wu studied at the National Central University of Nanjing in Nanjiang (南京市) (that university was renamed as Southeast University in 1988), and graduated with a BS in physics in 1934. Next, Wu pursued research at Zhejiang University (a future Nobel Laureate, Tsung-Dao Lee, was also a student there) and the Institute of Physics of the Academia Sinica. Wanting to further her studies, she emigrated from China to the United States, and enrolled at the University of California, Berkeley, to study physics. At the time, there was no post-doctorate program in physics in China. Initially, Wu wanted to attend the University of Michigan, but then switched to Berkeley because the student union at the University of Michigan only accepted men at the time. At Berkeley, Wu had the opportunity to work with Emilio Segrè, who later became a Nobel Prize recipient in 1959 for his discovery of the antiproton (Owen Chamberlain shared this prize with Segrè). Segrè and Wu later became very close friends and colleagues. In 1937, Wu was recommended for a fellowship, but sadly did not attain it. It was speculated that racial bias played a role. During the same year, Japan invaded China (the Sino-Japanese War), and Wu did not hear from her family who still lived in China. In fact, her departure to America was the last time that she ever saw her family in person. Despite such disappointments, Wu worked as a graduate student under the tutelage of Ernest O. Lawrence, who was a 1939 Nobel Prize recipient in physics for his invention of the cyclotron (a subatomic particle accelerator). In 1940, Wu received her doctorate from Berkeley.

While pursuing her Ph.D. at Berkeley, Wu met Luke Chia-Liu Yuan (袁家騮, a grandson of Shikai Yuan, 袁世凱, the first president of the new China after the fall of the Qing Dynasty), who was her classmate and fellow Chinese physicist. In 1942, the two married and eventually moved to the east coast, where Wu taught at Smith College while Yuan worked with radar devices at Princeton University. Then, due to wartime shortages of men in the country, Wu received different offers from Princeton, MIT, and Columbia. Wu accepted a non-research offer at Princeton to teach nuclear physics to naval officers. She was actually the very first female instructor who taught male students at Princeton.

During World War II when Wu was only in her 30's, her work in nuclear fission became well known, leading Wu to be selected to work on the Manhattan Project at the War Research Department at Columbia University in 1944. The Manhattan Project was a top-secret project that worked on the development of the atomic bomb for the Army. Her team developed instruments that could detect irradiation and devised a method of creating bomb-grade uranium, which involved enriching uranium ore so that uranium could be used as fuel for the bomb. The team was able to use a gaseous diffusion process to separate uranium into U-235 and U-238 isotopes, and also to improve existing Geiger counters, which are tools for measuring nuclear irradiation levels.

After the war, Wu remained at Columbia to carry on with research. It is interesting to note that although Wu's work in nuclear fission was becoming quite famous, it was presumed that she had not been promoted to be a faculty member presumably because she was an Asian woman and there was no woman teaching physics in a major American university at that time. Additionally, both Wu and her husband received offers from the National Central University in China, but they chose to remain in America due to the civil war and the communist victory in China. Furthermore, Wu finally heard that her family in China had survived the war. In 1947, Wu gave birth to their son Vincent Wei-Chen (緯承), who eventually studied nuclear science as well. Ultimately, Wu was promoted to associate professor in 1952.

Beginning in 1956, Wu started to collaborate with Tsung-Dao Lee of Columbia University and Chen Ning Yang of the Institute for Advanced Study in Princeton. Lee and Yang were theoretical physicists who, after conducting library research, hypothesized that the widely believed principle of parity (or simply put, of "symmetry"), which was a law of symmetry in physics, might not be true for weak nuclear forces such as that of beta decay. Consequently, Lee and Yang consulted Wu, who agreed to perform the experiment to either confirm or refute the widely held theory. Some responsibilities of carrying out such experiment included selecting the hardware, setting up the apparatuses, and planning the experimental procedure. This was a difficult experiment that required very precise and accurate planning. On January 15, 1957, while working with scientists Ernest Ambler, Raymond W. Hayward, Dale D. Hoppes, and Ralph P. Hudson from the US National Bureau of Standards in Washington, D.C., Wu and her group made a most astounding discovery in beta decay, confirming the theoretical predictions of fellow colleagues Lee and Yang. Wu's group recognized that instead of having both the right- and left-handed molecules behaving symmetrically in a single file, there exists a preferred, asymmetrical pathway of emission of weak subatomic particles. Here, Wu's group had used radioactive cobalt-60, a particle that decays via beta emission, to prove that the principle of parity (or of "symmetry") was indeed not applicable to weak interactions (such as beta decay). What resulted from this observation was the overturn of the principle of conservation of parity (or of "symmetry"), and news of this finding quickly spread. Soon after the news, fellow scientists from various laboratories confirmed the results from Wu's group through further experiments. The significance of this finding was not only important for the area of high energy physics, but it also aided the revision of the Standard Model theory.

[N.B. The following is a gist derived from McGrayne as well as from other sources: Beta decay is one of three varieties of radioactivity. In beta decay, a neutron inside a nucleus breaks apart, forming an electron, a proton and an electron antineutrino. In the variety of beta decay known as beta-minus, the exit of electrons and electron antineutrinos from the nucleus depletes the excess energy of the nucleus (i.e., releasing energy in the form of radioactivity). The original laws of parity and "symmetry" maintained that nuclei, atoms and molecules always functioned symmetrically. So it had long been expected that during beta decay, the exit of electrons from a nucleus would behave symmetrically (i.e., obeying the law of parity by releasing electrons symmetrically in multiple directions). However, Lee and Yang theorized that the exit of electrons from a nucleus during weak interactions (e.g., beta decay) might at times prefer one direction over others. In other words, the electrons might sometimes disobey the time-honored law of parity (or of "symmetry")].

The following is a more detailed description of what Wu and her National Bureau of Standards colleagues actually did. It is well known that one form of radioactive cobalt, cobalt-60 (^{60}Co), decays by the beta-minus variety to the stable isotope nickel-60 (^{60}Ni). During this decay, a neutron in the ^{60}Co nucleus disintegrates into a proton by emitting an electron and an electron antineutrino. By exposing ^{60}Co atoms to a temperature in the neighborhood of absolute zero and aligning them in a unifrom magnetic field at the same time, Wu and her colleagues found that ^{60}Co emitted more electrons in the direction opposite to that of the magnetic field. This finding was contrary to what one would expect if the law of parity had been observed. Under the experimental conditions of the study, the law of parity would require that the number of electrons emitted in a direction parallel to that of the magnetic field should be equal to that emitted in the opposite direction. Since the electrons in the experiment did not behave in this conventionally expected manner, Wu and her group succeeded in proving convincingly that beta-minus decay involving weak nuclear forces did not follow the law of parity.

Later in 1957, Lee and Yang were awarded the Nobel Prize for their theoretical contributions to this phenomenon. Unfortunately, Wu was left out of the Prize because her discovery was based on the ideas of her two colleagues.

After her most momentous breakthrough, Wu continued to pursue her love of research. For instance, in 1963, she confirmed the theory proposed by Richard P. Feynman and Murray Gell-Mann that the vector current in beta decay is conserved (i.e., "symmetrical"). Wu also spent some time studying the structure of hemoglobin.

Wu's three greatest scientific contributions can be summarized as: her precise beta decay experiments that became the basis for theoretical development during her early career; her confirmation of the parity violation theory; and eventually her confirmation of vector current conservation in beta decay. This last theory was coincidentally also proposed by two Nobel Prize winners. Many people strongly believed that Wu should have won a Nobel Prize for any of the aforementioned contributions.

Although Wu did not receive the Nobel Prize, she earned countless other prestigious awards for her research achievements, including the National Medal of Science (1975), the Bonner Prize (1975), the inaugural Wolf Prize in Physics (1978), and Columbia's Pupin Medal (1991), just to name a few. Moreover, she was elected as member in various prestigious professional organizations such as the National Academy of Sciences in 1958, the Royal Society of Edinburgh in 1969, and the American Academy of Arts and Sciences in 1972. Additionally, she was the first woman to receive the Research Corporation Award (1958), the first woman to be elected president of the American Physical Society (1976), the first woman to earn the Cyrus B. Comstock Award (1964), and the first woman to be presented with an honorary doctorate from Princeton University (1958). In 1965, Wu published her book, *Beta Decay*, which is still regarded as the standard reference in nuclear physics today. Another amazing feat was that Wu was the first living scientist to whom an asteroid was named after! On May 18, 1990, the Purple Mountain Observatory in Nanking named an asteroid that was discovered in 1965 after Wu. The asteroid is called the 2752 Wu Chien-Shiung.

In 1958, Wu became a full professor at Columbia. Furthermore, in 1973, she was appointed the very first Michael I. Pupin Professor of Physics at Columbia.

Wu continued her passion for nuclear research and teaching at Columbia until 1981 when she retired. Then, she spent her time lecturing and encouraging women to pursue scientific careers. She persistently advocated against gender discrimination in science. Besides being named "First Lady of Physics", Wu was also nicknamed "Chinese Marie Curie" and "Madame Wu".

On February 16, 1997, Wu passed away in New York, after her second stroke at the age of 84. Her grave is underneath the crepe myrtle trees grown by her father next to the Mingde School at Liuhe. Her husband donated all her belongings, including all her awards, to her alma mater, the Southeast University in Nanjiang. To honor her, the University has built the Chien-Shiung Wu Memorial Hall (http://chien-shiungwu.seu.edu.cn/), and all her belongings are now being displayed in that Hall.

When Wu first arrived in America in the 1930's, she was a woman from China who did not speak English very well and had to overcome many obstacles prevalent in the academic field at that time. Due to her perseverance, courage, and drive towards excellence, Wu became an icon in the field of physics. Professor Wu had an extraordinary life journey that inspires everyone. It is a shame that she was not awarded the Nobel Prize. However, Wu does not even need the Nobel Prize; she is already a legend who is much admired and revered by everyone, including many of the Nobel Prize winners. In 1986, four Nobel Prize Laureates — Chen Ning Yang, Tsung-Dao Lee, Samuel Chao Chung Ting and Yuan Tseh Lee — established the Wu Chien-Shiung Education Foundation in Taiwan to encourage the younger generations to strive for excellence in the sciences. Wu's legacy continues on today, and her contributions to the scientific community will live on for generations to come.

Finally, Suzanne Gould quoted Chien-Shiung, a gender discrimination fighter, well (Gould): "There is only one thing worse than coming home from the lab to a sink full of dirty dishes, and that is not going to the lab at all." She advocated for women to persist in pursuing careers in the sciences despite gender discrimination barriers.

Editors' Addition: Since females number more than 50% of humankind, this exemplary and fervent advocacy does carry an immense potential to change the world and is a shining example of Chien-Shiung's myriad legacies.

Editors' Note

A. Emilio Segrè, a Nobel Laureate, wrote in his 1980 book: "The trio of Chinese physicists (per editors: meaning T. D. Lee, C. N. Yang and Chien-Shiung Wu) shows what China's future contribution to physics could be if that country resumes its historic role as one of the leaders of civilization, as witnessed by the early European travelers, to their astonishment." From: Emilio Segrè. *From X-rays to Quarks — Modern Physicists and Their Discoveries*. Publishers: W.H. Freeman and Company, 1980, pp. 259–60.

B. As quoted by Sharon McGrayne, after the successful completion of her parity experiment, Wu commented, "These are moments of exaltation and ecstasy. A glimpse of this wonder can be reward of a lifetime" (McGrayne, page 276).
C. "Obsessed with physics" — Nobel Laureate Wolfgang Pauli about Chien-Shiung (Reynolds, page 112).
D. Chien-Shiung once said: "Beta decay was .... like a dear old friend. There would always be a special place in my heart reserved especially for it. "(Newman and Ysilantis, pp. 390–1).
E. "Chien-Shiung did not have everything that she needed to thrive. She faced discrimination because of her gender and race. Nevertheless, she was able to succeed because of her determination, hard work, and bravery" (Cooperman, page 98).
F. As quoted by Noemie Benczer-Koller, one of Wu's students, Wu believed that, "It is the courage to doubt what has long been believed and the incessant search for verification and proof that push the wheels of science forward." In addition, Noemie Benczer-Koller asserted that "Her remarkable life can be portrayed by an ancient Chinese poem by Qu Yuan (ca. 340–278 BCE)": "Although the road is long and arduous, I am determined to explore its entire length" (Benczer-Koller, pp. 13 and 16).
G. Nobel Laureate Tsung-Dao Lee once commended: "C. S. Wu was one of the giants of physics. In the field of beta-decay, she had no equal." (Famed Physicist, Chien-Shiung Wu dies at 84, Columbia University.)
H. "She (per the editors: i.e., Chien-Shiung) had pioneered one of the most important experiments in physics. Of course, she could not have done it without her coworkers at the NBS (per the editors: i.e., the National Bureau of Standards), but it was her idea and she pushed it through." (Hammond, page 88).

Honors, Awards, and Distinctions (Wikipedia)

- Elected a fellow of the American Physical Society (1948)
- Elected a member of the U.S. National Academy of Sciences (1958)
- First woman with an honorary doctorate from Princeton University (1958)
- Achievement Award, American Association of University Women (1959)
- Research Corporation Award (1959)
- John Price Wetherill Medal, The Franklin Institute (1962)
- American Association of University Women Woman of the Year Award (1962)
- Comstock Prize in Physics, National Academy of Sciences (1964)
- Chi-Tsin Achievement Award, Chi-Tsin Culture Foundation (1965)
- Honorary Fellow Royal Society of Edinburgh (1969)
- Scientist of the Year Award, Industrial Research Magazine (1974)

- Tom W. Bonner Prize, American Physical Society (1975)
- First female President of the American Physical Society (1975)
- National Medal of Science (1975)
- First person selected to receive the Wolf Prize in Physics (1978)
- Ellis Island Medal of Honor (1986)
- First living scientist to have an asteroid (2752 Wu Chien-Shiung) named after her (1990)
- Pupin Medal, Columbia University (1991)
- Elected one of the first foreign academicians of the Chinese Academy of Sciences (1994)
- Inducted into the National Women's Hall of Fame (1998)

References and Recommended Readings

1. Wu CS, Ambler E, Hayward W, Hoppes DD, Hudson RP. (1957) Experimental test of parity conservation in beta decay. *Phys Rev* **105**: 1413–15.
2. Lee TD, Yang CN. (1956) Question of parity conservation in weak interactions. *Phys Rev* **104**: 254–8.
3. Lee TD, Yang CN. (1956) Elementary particles and weak interactions. Brookhaven National Laboratory, Associated Universities Inc., under contract with the United States Atomic Energy Commission. Available from the Office of Technical Services, Department of Commerce, Washington 25 D.C., USA.
4. Wu CS, Moszkowski SA. (1966) Beta decay. Publishers: Interscience Publishers, Geneva, Switzerland.
5. Wu CS. (1977) Subtleties and surprises: The contribution of beta decay to an understanding of the weak interaction. *Ann NY Acad Sci*, 294, Issue 1, pp. 37–51. https://doi.org/10.1111/j.1749-6632.1977.tb26471.x.
6. Wu CS. (1983) The discovery of the parity violation in weak interactions and its recent developments. Nishina Memorial Foundation, Bunkgo-ku, Tokyo, Japan.
7. Kass-Simon G, Farnes P, Nash D. (1989) Women of Science: Righting the Record. Indiana University Press, Bloomington, Indiana.
8. McGrayne SB. (1993) Chapter on Chien-Shiung Wu. In: *Nobel Prize Women in Science: Their Lives, Struggles and Momentous Discoveries*. A Birch Lane Press Book, Published by Carol Publishing Group, New York, NY.
9. Reynolds MD. (1999) Chapter on Chien-Shiung Wu — Experimental Nuclear Physicist. In *American Women Scientists: 23 Inspiring Biographies, 1900–2000*. Publishers: McFarland & Company, Inc., Jefferson, North Carolina, pp. 112–8.
10. From Madame Wu's first-hand account of her experimental determination of parity violation which was given to the International Conference on the History of

Original Ideas and Basic Discoveries, held in Erice, Sicily, July 27–August 4, 1994; vide: *History of Original Ideas and Basic Discoveries in Particle Physics*, edited by H.B. Newman and T. Ypsilantis, Plenum Press, New York, N.Y., 1996, pp. 390–1.

11. Cooperman SH. (2004) *Chien-Shiung Wu: Poineering Physicist and Atomic Researcher* (Women Hall of Famers in Mathematics and Science). Publishers: The Rosen Publishing Group, Inc., New York, NY.
12. Benczer-Koller N. (2009) *Chien-Shiung Wu 1912–1997: A Biographical Memoir*. National Academy of Sciences, Washington DC.
13. Famed Physicist Chien-Shiung Wu Dies at 84. Columbia University. www.columbia.edu/cu/record/archives/ vol22/vol22_iss15/record2215.16.html
14. Hammond R. (2009) *Chien-Shiung Wu: Pioneering Nuclear Physicist* (Makers of Modern Science). Chelsea House Pub (L), New York, NY.
15. Wheeler JC. (2012) *Chien-Shiung Wu: Phenomenal Physicist*. Checkerboard Biography Library: Women in Science. Publishers: Checkerboard Books, New York, NY.
16. Chiang Tsai-Chien. (2013) Translated by Wong Tang-Fong. *Madame Wu Chien-Shiung: The First Lady of Physics Research*. World Scientific Publishing, Co., Singapore.
17. Gould S. (2013) "This Brilliant Female Physicist was Overlooked for a Nobel Prize." Blog on American Association of University Women (AAUW) website, September 11, 2013 (accessed on December 1, 2014).
18. http://c250.columbia.edu/c250_celebrates/remarkable_columbians/chien-shiung_wu.html.
19. http://www.nwhm.org/education-resources/biography/biographies/chien-shiung-wu/.
20. http://womenshistory.about.com/od/sciencephysics/a/Chien-Shiung-Wu.html.
21. http://bnrc.berkeley.edu/Famous-Women-in-Physical-Sciences-and-Engineering/chien-shiung-wu.html.
22. Chien-Shiung Wu. Wikipedia, accessed on April 7, 2017.
23. Wu experiment. Wikipedia, accessed on April 7, 2017.
24. Beta decay. Wikipedia, accessed on April 7, 2017.
25. Chien-Shiung Wu Memorial Hall. http://chien-shiungwu.seu.edu.cn/(Memorial Hall in honor of Laureate Wu located at the Southeast University in Nanjing).
26. Wu, Chien-shiung. The Asian American Encyclopedia. Editor: Franklin Ng. Publisher: Marshall Cavendish Corp., 1995, p. 1686.

Photo 15.1. Chien-Shiung with father Wu Zong-Yi, mother Fan Fu-Hua, and brother Wu Chien-Ying. (From Ref. 16; courtesy of World Scientific Publishing Co.)

Photo 15.2. Chien-Shiung with uncle Wu Zhuo-Zhi who financed her trip to pursue graduate studies in the US. (From Ref. 16; courtesy of World Scientific Publishing Co.)

Photo 15.3. Chien-Shiung seen here with J. Robert Oppenheimer (middle) and Emilio Segrè (left) in 1940 at the University of California at Berkeley. (From Ref. 16; courtesy of the World Scientific Publishing Co.)

Photo 15.4. Chien-Shiung working with her electronic equipments in the Laboratory of Columbia University. (Courtesy of the US Library of Congress.)

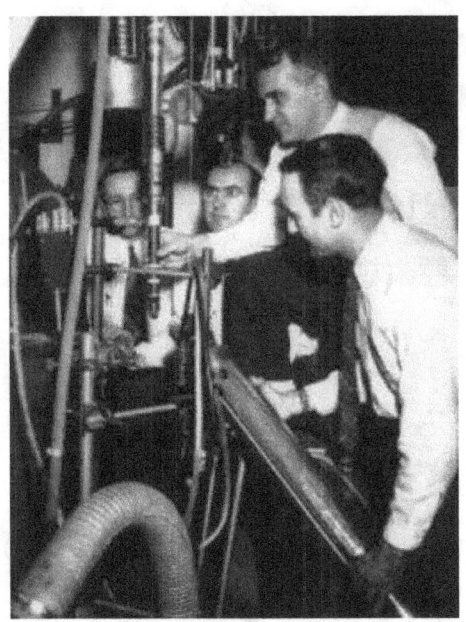

Photo 15.5. Chien-Shiung's scientific colleagues (from left to right) Raymond W. Hayward, Dale D. Hoppes, Ernest Ambler and Ralph P. Hudson in front of the exemplary experimental setup. (Courtesy of the American Physical Society.)

C- S Wu, E. Ambler, R.W. Hayward, D.D. Hoppes, and R.P. Hudson

National Bureau of Standards, Washington, D.C.
Discovering Parity Isn't

In 1956, the National Institute of Standards (NIST) was the National Bureau of Standards (NBS), and just as active in experimenting with fundamental physical processes.
In the early twentieth century, physicists assumed that our world is indistinguishable from its mirror image, an idea called "parity conservation." At NBS during that slow period between Christmas and New Year's, physicists Chien-Shiung Wu, Ernest Ambler, Raymond W. Hayward, Dale D. Hoppes and Ralph P. Hudson discovered quite convincingly that our world is distinguishable from its mirror image. The old campus of the National Bureau of Standards is now home to the University of the District of Columbia, in Washington, D.C. On November 9, 2011, Kate Kirby, APS's Executive Officer presented a commemorative plaque recognizing the UDC as an American Physical Society Historic Site, to Allen L. Sessoms, UDC President. Dr. Sessoms is a nuclear physicist and an APS Fellow (2008).

Pictured left to right: Isadora Posey, Chair of the Department of Chemistry & Physics at UDC; Kafayat Olayinka, physics major at UDC; Alan Chodos, Associate Executive Officer of APS; Beverly Hartline, Dean of Graduate Studies at UDC; Katharine Gebbie, Director of the Physical Measurement Laboratory at NIST; Allen Sessoms, President of UDC; Kate Kirby, Executive Officer of APS; Ben Bederson, Chair of the APS Historic Sites Committee; LaVonne Manning, Associate Professor, UDC Computer Science and Information Technology; Patricia Thorstenson, UDC Professor of Chemistry.

Allen Sessoms, President of UDC, and Kate Kirby, Executive Officer of APS, holding the bronze plaque.
Plaque Inscription
In 1956, at this site, which was then on the campus of the National Bureau of Standards, physicists C-S Wu, E. Ambler, R.W. Hayward, D.D. Hoppes and R.P. Hudson performed an experiment which revealed that in certain nuclear processes pairs of events that are merely mirror images of each other occur with different probabilities. This discovery revolutionized our understanding of nature's fundamental laws.
HISTORIC PHYSICS SITE, REGISTER OF HISTORIC SITES
AMERICAN PHYSICAL SOCIETY

Photo 15.6. Register of Historic Physics Site: "Parity is not conserved" Experiment. (Photo: Alan Etter/UDC. Reprinted with permission of the American Physical Society.)

1959

Dr. Wu Wins Achievement Award

A woman scientist whose experiments disproving the law of parity have been termed "the solution to the number one riddle of atomic and nuclear physics" received the 1959 AAUW Achievement Award, which carries a stipend of $2500, Tuesday evening of Convention Week. Dr. Chien-Shiung Wu, Professor of Physics at Columbia University, was cited for her "search which has led to the solution of mysteries on the fringe of human knowledge" by Dr. Janet Howell Clark, Consultant to the AAUW Fellowships Awards Committee, who went on to describe the contributions and discoveries that had earned the awardee the right to be called "the world's foremost woman in experimental physics."

an experiment to prove or disprove this law and suggested several possible tests, people were very skeptical about it, so firm was the belief of the great physicists, and one eminent authority wrote to another on January 17, 1957:

I do not believe the law of a weak left hand, I am ready to bet a high sum that the present experiments will give symmetrical results.

"Now you know that scientist was playing a sure losing game, because the nonconservation of parity was officially announced one day earlier in the *New York Times*, on January 16, 1957. But you can well imagine the nightmare which my co-workers and I lived through in those two weeks prior to the official announcement

Photo 15.7. Chien-Shiung accepting the 1959 American Association of University Women (AAUW) Achievement Award, an event detailed in the October 1959 issue of the "AAUW Journal". (Courtesy of AAUW Archives, Washington, DC.)

Photo 15.8. Chien-Shiung and Luke greeting Chou En-Lai in China in early 1973. (From Ref. 16; courtesy of World Scientific Publishing Co.)

Photo 15.9. Chien-Shiung and Luke met with Deng Xiao-Ping in China in the 1980s, along with Chen Ning Yang (second left) and Samuel Chao Chung Ting (third left). (From Ref. 16; courtesy of World Scientific Publishing Co.)

Photo 15.10. Chien-Shiung receiving the National Medal of Science from the US President Gerald R. Ford in 1975. (Courtesy of The White House.)

Photo 15.11. Chien-Shiung was awarded the inaugural Wolf Prize of Israel in 1978, with the Prime Minister Menachem Begin (third from the left) honoring her. (From Ref. 16; courtesy of World Scientific Publishing Co.)

Photo 15.12. Chien-Shiung with her family in front of their New York residence. From left to right: son, Vincent Wei-Chen; Laureate Wu; granddaughter, Jada; daughter-in-law, Lucy Lyon; and husband, Luke. (From Ref. 16; courtesy of World Scientific Publishing Co.)

Appendix A

The Nobel Prize

Susie Q. Lew and Keith K. Lau

The Nobel Prize consists of a series of international honors bestowed annually since 1901 for outstanding work in physics, chemistry, literature, peace, and physiology or medicine. Since 1969, the "Sveriges Riksbank Prize in Economic Sciences in Memory of Alfred Nobel" has generally received concurrent recognition as the (sixth) Nobel (for Economics).

Alfred Nobel (1833–1896) invented dynamite, held 355 patents, and owned Bofors AB, a Swedish arms manufacturer. A Swedish national, he was a chemist, inventor, engineer, entrepreneur, business man, author, and pacifist. His will left 31 million SEK (approximately $2 million US dollars at that time or about $0.5 billion today) to fund the prizes "for the Greatest Benefit to Mankind."

The Nobel Foundation in Stockholm, Sweden, manages the finances and administration of the prizes. It has no involvement in the selection of the candidates or recipients. The Nobel Prizes are traditionally announced on successive days in October.

In 1968, Sveriges Riksbank (Sweden's central bank) made a major donation to the Nobel Foundation. In honor of the bank's 300th Anniversary, it established The Sveriges Riksbank Prize in Economic Sciences in Memory of Alfred Nobel.

The Royal Swedish Academy of Sciences selects the Nobel Laureates in Physics, Chemistry, and Economics. The Nobel Assembly, a committee at the Karolinska Institute, selects the Nobel Laureates in Physiology or Medicine. The Swedish Academy selects the Nobel Laureates in Literature. The Norwegian Nobel Committee, appointed by the Storting (the Parliament of Norway), selects the Nobel laureates in Peace.

The Award Ceremony

Each Prize includes a medal, a personal diploma, and a cash award. A person or organization awarded the Nobel Prize is called Nobel laureate. The word "laureate" derives from the laurel wreath. The custom of honoring special achievement with a wreath-of-honor comes from ancient Greece.

The amount of prize money varies depending on the Nobel Foundation's income each year. In recent years, the cash award averages about one million US dollars. The recipients generally donate their prize money to benefit scientific, cultural, or humanitarian causes.

The annual Nobel Prize Award Ceremony (except for the Peace Prize) takes place at the Stockholm Concert Hall in Stockholm, Sweden. Presentation speeches extoll the Nobel laureates and their accomplishments. His Majesty the King of Sweden presents each Laureate with a diploma and a medal.

The Nobel Banquet immediately follows the ceremony. It is held in the Blue Hall of the Stockholm City Hall. The first banquet in 1901 was attended by 113 male guests. Today, some 1,300 guests are invited. They now include academics, government officials, cultural icons, industrialists, diplomatic representatives, and the Royal Family of Sweden. During the early years, guests were seated at tables arranged in the shape of a horseshoe. Today, long tables surround the table of honor set in the middle of the banquet hall.

The Nobel Prize Award Ceremony in Stockholm and the Nobel Banquet that follows are formal affairs. Gentlemen wear white tie and tails, while ladies dress in evening gowns. Wearing national costume is an alternative to white tie or evening gown. The Queen's gown, an item of special interest, lends a dash of color to the otherwise somber stage at the Stockholm City Hall and male participants wearing their white tie and tails.

Separately, the Chairman of the Norwegian Nobel Committee presents the Nobel Peace Prize to the Laureate in Oslo, Norway. Their Majesties the King and Queen of Norway, the Norwegian government, Storting representatives and an invited audience attend the ceremony. The Nobel Peace Prize Banquet is held in the Oslo Grand Hotel after the award ceremony. Attendance includes the King and Queen of Norway, President of the Storting, Prime Minister, and about 250 distinguished guests.

Some Interesting Facts about the Nobel Prize

"A prize amount may be equally divided between two works, each of which is considered to merit a prize. If a work that is being rewarded has been produced by two or three persons, the prize shall be awarded to them jointly. In no case may a prize amount be divided between more than three persons."

Alfred Nobel left his fortune to finance annual prizes to be awarded "to those who, during the preceding year, shall have conferred the greatest benefit on mankind." The prize is not awarded posthumously. However, if a person is awarded a prize and dies before receiving it, the prize may still be presented. As of 2017, 584 prizes have been awarded to 923 laureates. "If none of the works under consideration is found to be of the importance indicated in the first paragraph, the prize money shall be reserved until the following year. If even then, the prize cannot be awarded, the amount shall be added to the Foundation's restricted funds."

The Nobel Prize remains widely regarded as the most prestigious awards given for intellectual achievement in the world.

Acknowledgments

Photos A.1 to A.7 below; courtesy of the Nobel Foundation.

References

https://www.nobelprize.org Accessed July 8, 2018.
https://en.wikipedia.org/wiki/Nobel_Prize Accessed July 8, 2018.

Photo A.1. Photo of Alfred Nobel. (© ® The Nobel Foundation.)

Photo A.2. The Nobel Medal for Physics and Chemistry.
The medal of the Royal Swedish Academy of Sciences represents Nature in the form of a goddess resembling Isis, emerging from the clouds and holding in her arms a cornucopia. The veil which covers her cold and austere face is held up the Genius of Science. (© ® The Nobel Foundation.)

Photo A.3. The Nobel Medal for Physiology or Medicine.
The medal of the Nobel Assembly at the Karolinska Institute represents the Genius of Medicine holding an open book on her lap, collecting the water pouring out from a rock in order to quench a sick girl's thirst. (© ® The Nobel Foundation.)

Photo A.4. The Nobel Medal for Literature.
The medal of the Swedish Academy represents a young man sitting under a laurel tree who, enchanted, listens to and writes down the song of the Muse.
The inscription on the medals read: *Inventas vitam iuvat excoluisse per artes* translated "inventions enhances life which is beautified through art."
The name of the Nobel laureate is engraved on the plate below the figures.
(© ® The Nobel Foundation.)

(a)   (b)

Photo A.5. The Nobel Peace Prize Medal.

(a). *Front View*: Alfred Nobel.

(b). *Back View*: A group of three men forming a fraternal bond.

The inscription on the medal reads: *Pro pace et fraternitate gentium* translated "For the peace and brotherhood of men."

"Prix Nobel de la Paix", the relevant year, and the name of the Nobel Peace Prize Laureate are engraved on the edge of the medal.

(© ® The Nobel Foundation.)

(a)   (b)

Photo A.6. The Medal for the Sveriges Riksbank Prize in Economic Sciences in Memory of Alfred Nobel.

(a). *Front View*.

The upper half depicts Alfred Nobel.

The inscription on the medal reads: Sveriges Riksbank till Alfred Nobels Minne 1968 translated The Sveriges Riksbank, in memory of Alfred Nobel, 1968.

The lower half displays the bank's crossed horns of plenty.

(b). *Back View*.

The medal for The Sveriges Riskbank Prize in Economic Sciences in Memory of Alfred Nobel shows the North Star emblem of the Royal Swedish Academy of Sciences, dating from 1815, with the words "Kungliga Vetenskaps Akademien" (The Royal Swedish Academy of Sciences) around the edge.

The name of the Economics Laureate is engraved on the edge of the medal.

(© ® The Nobel Foundation.)

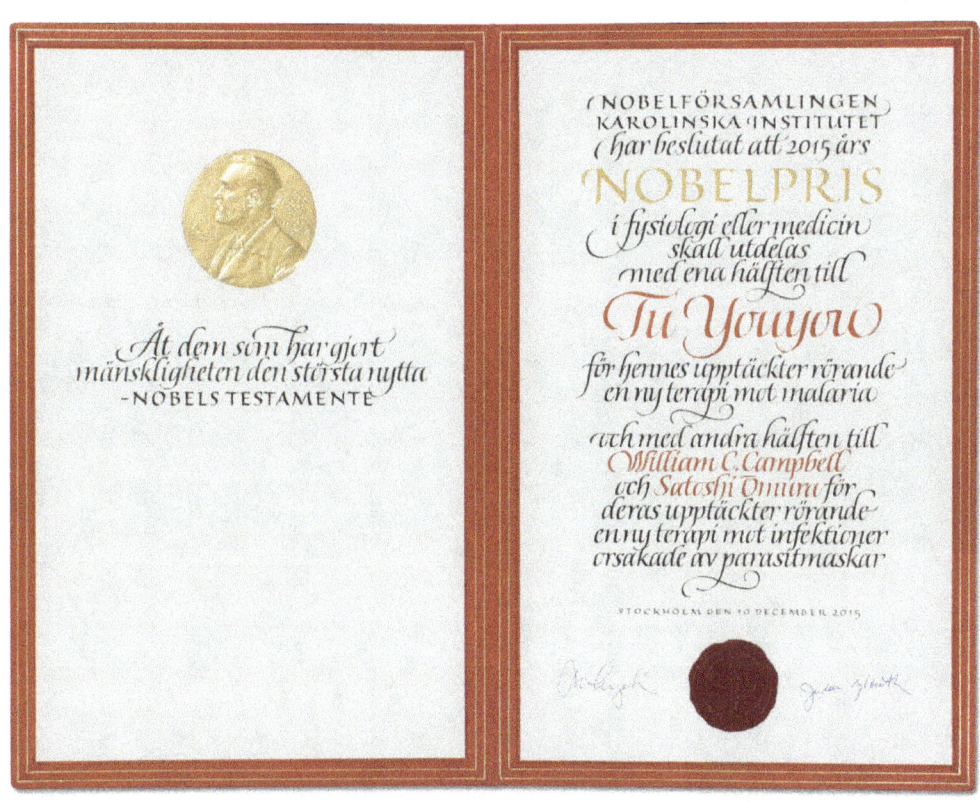

Photo A.7. A sample of the Nobel Diploma; that of Laureate Youyou Tu's. (Copyright of The Nobel Foundation 2015. Calligrapher: Susan Duvnäs; Book binder: Ingemar Dackéus; Photo reproduction: Lovisa Engblom.) (© ® The Nobel Foundation.)

Appendix B

The Lasker Awards

Susie Q. Lew and Hon-Lok Tang

The Lasker Awards recognize the contributions of researchers, clinician scientists, and public servants who have made major advances in the understanding, diagnosis, treatment, cure, or prevention of disease. Albert and Mary Lasker created the Lasker Awards program in 1945 to highlight fundamental biological discoveries and clinical advances that improve health, and to draw attention to the importance of public support of science.

Lasker Awards are currently given in four categories.

- The Albert Lasker Basic Medical Research Award recognizes a fundamental discovery in medicine or human physiology.
- The Lasker-DeBakey Clinical Medical Research Award honors outstanding work for the understanding, diagnosis, prevention, treatment, and cure of disease.
- The Lasker-Koshland Special Achievement Award in Medical Science identifies research accomplishments and scientific statesmanship that engender the deepest feelings of awe and respect. It is not awarded every year.
- The Lasker-Bloomberg Public Service Award pays homage to an individual or organization that improves the public's understanding of medical research, public health, or healthcare; providing financial support for initiatives to enhance public health and/or medical research; or serving as a spokesperson for medical research or public health; providing outstanding public health practice.

The Albert and Mary Lasker Foundation, endowed by the estate of Albert and Mary Lasker, also receives gifts from donors. Three awards bear names of subsequent donors: Michael R. Bloomberg represented by the Bloomberg Philanthropies, Michael E. DeBakey represented by the DeBakey Medical Foundation, and Daniel E. Koshland Jr.

The Lasker Awards are frequently referred to as "America's Nobels." They are considered the most prestigious American award in medical research. The Lasker Award has the reputation for identifying future winners of the Nobel Prize. Eighty-six Lasker laureates have received the Nobel Prize. Almost 50% of the Albert Lasker Awardees for Basic Medical Research have also received a Nobel Prize. The Lasker Awards carry an honorarium of $250,000. The recipient also receives a statuette.

Albert Lasker

Albert Davis Lasker (1880–1952) amassed his fortune as an executive at Lord & Thomas, a Chicago advertising agency. He made the firm the largest advertising agency in the

United States. Lasker was known as the "father of modern advertising" with copywriting that appealed directly to the psychology of the consumer.

He had other business interests as well. Lasker owned the Chicago Cubs baseball team and was instrumental in moving the Cubs to Wrigley Field. His interest in baseball led to the creation of the office of the Commissioner of Baseball to reform baseball's governing authority. He built the Lasker Golf Course (ranked 23 in the world) in Lake Forest, Illinois, and later donated the entire property to the University of Chicago. In 1921, President Harding appointed Lasker as chairman of the United States Shipping Board, where he served for two years.

He had three children, Mary Lasker Block (1904), Edward (1912), and Frances Lasker Brody (1916) with his first wife Flora Warner (1880–1934). His subsequent marriages to Doris Kenyon (married in 1938), an actress, and Mary Woodard Reinhardt (married 1940), an industrial designer; were without issue. Albert died (1952) in New York at the age of 72 from colon cancer.

Mary Lasker

Mary Lasker (1900–1994) was an American health activist and philanthropist. She was born in Wisconsin and attended the University of Wisconsin. She graduated from Radcliffe College in Cambridge, Massachusetts, majoring in Art History. Mary Woodard married Paul Reinhart of Reinhardt Galleries in New York City. After their divorce, she created a fabric company: Hollywood Patterns. In 1938, she became president of the Birth Control Federation of America, the precursor of the Planned Parenthood Federations. She married Albert Lasker in 1940.

Albert and Mary Lasker were nationally prominent philanthropists and formed the Albert and Mary Lasker Foundation to execute their philanthropy. Mary served as president of the Foundation. She also served as director, chairperson or trustee of the American Cancer Society, the United Cerebral Palsy Research and Education Foundation, the National Committee for Mental Hygiene, and a range of other medical and cultural organizations. Mary Lasker's achievements were recognized with the Presidential Medal of Freedom (1969), the Four Freedoms Award (1987), the Congressional Gold Medal (1989), and the Albert Schweitzer Gold Medal for Humanitarianism (1992). In 1971, she became the second female member of the Board of Director of Braniff Airways, Incorporated; a rare occurrence at the time. In 2009, the United States Postal Service honored Mary Lasker with the issuance of a 78-cent stamp. The latter recognizes a renewed US government commitment to fund biomedical research. She died in 1994 at the age of 93, leaving more than $10 million to the Lasker Foundation to support medical research and urban beautification.

Albert and Mary Lasker's Public Service

Albert Lasker pursued his passion for philanthropy after his retirement. He aggressively promoted the expansion of medical research in the United States. The Albert and Mary Lasker Foundation, located in New York City, was established in 1942 to support medical research and to create the Lasker Awards.

References

www.laskerfoundation.org Accessed July 8, 2018.
https://en.wikipedia.org/wiki/Albert_Lasker Accessed July 8, 2018.
www.newworldencyclopedia.org/entry/Albert_Lasker Accessed July 8, 2018.
https://en.wikipedia.org/wiki/Mary_Lasker Accessed July 8, 2018.

Photo B.1. Laureate Youyou Tu's 2011 Lasker-DeBakey Clinical Medical Research Award statuette and certificate. (Photo taken by Dr. Keith K. Lau at the China Academy of Chinese Medical Sciences building in Beijing.)

Photo B.2. The Lasker *Winged Victory* statuette. (Courtesy of the Albert and Mary Lasker Foundation.)

The original *Winged Victory of Samothrace* was created by the Greeks in the period between 190–180 BC and is considered one of the Louvre's three greatest masterpieces, together with Leonardo da Vinci's *Mona Lisa* and the *Venus de Milo* sculpture. *Winged Victory*, which is eight feet tall, portrays the Greek Goddess of Victory standing on the prow of a ship with her wings spread and her clinging garments rippling in the wind as she descends from the sky to celebrate the naval triumph of the fleet. In creating the Lasker Awards, Mary Lasker conceived and designed the *Winged Victory* statuette, which is one foot tall, to symbolize a body of creative biomedical research that produces victory over disability, disease, and death. (Goldstein JL. 60 years of winged victories for biomedical research. *Nature Medicine* 2005; **11**(10), pp ii–iv.)

Photo B.3. Mary and Albert Lasker. (Courtesy of the Albert and Mary Lasker Foundation.)

"悟已往之不諫, 知來者之可追。實迷途其未遠, 覺今是而昨非"。

來自
歸去來辭
作者: 陶淵明, (公元 365–427), 六朝文豪

"Aware of past mistakes, cognizant of the fact that a better future is within reach. Realizing that one has not completely lost one's way, acknowledging that today is right and yesterday, wrong."

From Returning Speech

by Tao Yuanming, 365–427 CE, a renowned scholar in the Six Dynasties Period.

Editors' Note: One should not be discouraged by setbacks.

Index

A

Academy of Chinese Medicine, 167
accomplishments, Ting, Samuel Chao Chung, 44–45
achievements, Li, Choh Hao, 188–189
ACT. *See* artemisinin combination therapy
actinomycin D, 202, 209
adrenocorticotrophic hormone (ACTH), 187
Adrian, Richard, 120
Adventures of Tom Sawyer, The (Twain), 21
Aequorea victoria, 120
Akers-Jones, David, 52
Albert and Mary Lasker Foundation, 272, 274
Albert Lasker Basic Medical Research Award, 272
Albert Lasker Clinical Medical Research Award, 202
α-globin gene, 229
α-globin protein, 229
α-thalassemia, 229
Alpha Magnetic Spectrometer (AMS), 42, 43
Alzheimer's disease, 137
Ambler, Ernest, 249
American Society of Hematology, 231
"America's Nobels". *See* Lasker Awards
amino acid, 186, 187
amplitude, 134
AMS. *See* Alpha Magnetic Spectrometer
AMS-01, 42
AMS-02, 42, 43
anopheles mosquito, 168
Antimalarial Project 523, 168
"Anti-Rightist Movement", 149
Area of Darkness, An (Naipaul), 154
aristolochic acid nephropathy, 171
Artemisia annua L., 169
artemisinin, 169–171
artemisinin combination therapy (ACT), 169
art films, Gao, Xingjian, 106
Ashkin, Art, 75
astrophysics research, Lee, Tsung-Dao, 23
atom interferometers, Chu, Steven, 75
Augustana College, 89
Aurora Biosciences Corporation, 122
avermectin, 167
avid reader, Lee, Tsung-Dao as, 21
Award Ceremony, 264, 265
awards
    Chu, Steven, 76
    Kan, Yuet Wai, 230–231
    Lee, Yuan Tseh, 61
    Li, Choh Hao, 190
    Li, Min Chiu, 210
    Mo, Yan, 154
    Ting, Samuel Chao Chung, 46
    Tsien, Roger Yonchien, 122
    Tsui, Daniel Chee, 90
    Tu, Youyou, 172, 173
    Wu, Chien-Shiung, 250, 252, 253
    Yang, Chen Ning, 9

B

Ba, Jin, 103
Bank of East Asia, 228
Bardeen, J., 26
BCS theory for superconductivity, 26
Begin, Menachem, 261
Beijing Normal University, 149
Bell Laboratories, 75, 90
Bellow, Saul, 50, 112
Bernstein, Jeremy, 24, 25
Bernstein, R. B., 60
Bessel functions, 134
beta decay, 249, 250
Beta Decay (Wu), 250
β-endorphin, 185, 187, 190, 191
beta-minus, 249
β-thalassemia, 229
Big Breasts and Wide Hips (Mo), 150, 152, 153
Bishop, J. Michael, 230
Blake, William, 8
Bloomberg, Michael R., 272
Blumberg, Baruch S., 50
Bollinger, Lee, 28
books, Mo, Yan, 155, 156
Bovet, D., 34
Boxer Indemnity Scholarship, 118

Boyer, Herbert, 230
Bragg, W. L., 21
Bush, George H. W., 119
Bush, George W., 128

C

Caenorhabditis elegans, 120
Cambridge Electron Accelerator, 41
Campbell, William C., 167
Camus, A., 34
cancer therapy, Li, Min Chiu, 202, 204, 208, 209
Carbone, Paul, 211
career
    Chu, Steven, 73
    Gao, Xingjian, 103
    Kao, Charles Kuen, 133, 136
    Li, Choh Hao, 188
cDNA, 120, 121
Chalfie, Martin, 120–122
Chandrasekhar, S., 23
Chang, Ming-Chang, 21
Chang, Stephen, 229
Charles K. Kao Foundation, 137
charm quark, 44
chemotherapeutic agents, 211
chemotherapy for malignancy, Li, Min Chiu, 202–211
Chen, Anna Gustafsson, 154
Chen, K. K., 170
Chen, Zhu, 167
Chiang, Kai-Shek, 149
"China Dream", 7
China–US Physics Examination Program (CUSPEA), 27
Chinese Democracy Movement, 104
Chinese herbs nephropathy, 171
Chinese University of Hong Kong (CUHK), 7–9, 136, 190, 231
chlorambucil, 209
chloroquine, 168
Choa, Gerald, 229
Choh Hao Li Memorial Lectureship, 189
Chou, En-Lai, 259
choriocarcinoma treatment, Li, Min Chiu, 202, 203, 209–212
choriogonadotropin, 203–205

Christie, Dugald, 203
Chu, E., 211
Chu, Ju-Chin, 73
Chu, Paul, 8
Chu, Steven, 214
    approach, 75
    Art, Ashkin, 75
    atom interferometers, 75
    awards and recognitions, 76
    Bell Labs, 75
    brilliant research scientist, 74
    career, 73
    early life, 73
    education, 73, 74
    family, 73
    graduate program in physics, 74
    "hands-on" training, 74
    honorary degree, 76
    laser cooling application, 75
    Nobel biography, 74
    Nobel Prize motivation, 73–76
    outstanding features, 74
    television interview (2004), 74
Cinchona ledgeriana, 170
cobalt-60, 249, 250
Cohen-Tannoudji, Claude, 73, 75
Cole, D. R., 188
collisions of particles, 41, 42
colorectal cancer study, Li, Min Chiu, 209
Commins, Eugene, 75
Communist Party, 149, 151
Condit, Paul, 205
conservation of parity, 23
corticotropin, 187
Crawford, David, 218
Cronin, James, W., 51
crossed molecular beam approach, 60
CUHK. *See* Chinese University of Hong Kong
Cultural Revolution, 103, 149, 150
CUSPEA. *See* China–US Physics Examination Program

D

DeBakey, Michael E., 272
Deng, Jiaxian, 6, 7
Deng, Xiaoping, 149, 260

DeVita, V. T., 211–213
dihydroartemisinin, 169
Dirac, P. A. M., 26
discovery
    beta decay, Wu, Chien-Shiung, 249, 250
    DNA polymorphisms, Kan, Yuet Wai, 228–231
    green fluorescent protein, Tsien, Roger Yonchien, 118–122
    J/Psi particle, Ting, Samuel Chao Chung, 40–44
    Kao, Charles Kuen, 134, 135
    qinghaosu (artemisinin), Tu, Youyou, 170
DNA polymorphism, 228–231
Doctor of Science degree, Yang, Chen Ning, 8
Dunne, Brendan, iii
Dutrait, Liliane, 103
Dutrait, Noël, 103
dynamics of chemical elementary processes, Lee, Yuan Tseh, 58–61

E
early life
    Chu, Steven, 73
    Kan, Yuet Wai, 228, 229
    Kao, Charles Kuen, 133
    Lee, Tsung-Dao, 21
    Lee, Yuan Tseh, 58, 59
    Li, Choh Hao, 185, 189
    Ting, Samuel Chao Chung, 40, 41
    Tsien, Roger Yonchien, 118
    Tsui, Daniel Chee, 87, 88
    Tu, Youyou, 167, 168
    Wu, Chien-Shiung, 247
    Yang, Chen Ning, 5
Eddington, Arthur, 21
education
    Chu, Steven, 73, 74
    Gao, Xingjian, 103
    Kan, Yuet Wai, 228
    Kao, Charles Kuen, 133
    Lee, Tsung-Dao, 21, 22
    Lee, Yuan Tseh, 59
    Li, Choh Hao, 185, 189
    Mo, Yan, 152
    Ting, Samuel Chao Chung, 40, 41
    Tsien, Roger Yonchien, 118, 119
    Tu, Youyou, 167, 168
    Wu, Chien-Shiung, 247, 248
    Yang, Chen Ning, 5
Einstein, Albert, 42
electromagnetic force, 24
electromagnetic waves, 134
Elementary Particles (Fermi), 23
enviornment, Tu, Youyou, 167, 168
Escherichia coli, 120
Evans, Herbert M., 186
Evans, W. V., 185
Expanding Universe: Astronomy's "Great Debate", The (Eddington), 21
experimental human subject, Tu, Youyou, 169

F
family
    Chu, Steven, 73
    Kan, Yuet Wai, 228
    Kao, Charles Kuen, 133
    Lee, Yuan Tseh, 61
    Li, Choh Hao, 185, 189
    Mo, Yan, 149
    Tu, Youyou, 167
    Wu, Chien-Shiung, 247
    Yang, Chen Ning, 5
Fan, Fuhua, 247
Faulkner, William, 149, 150
female–male relationships, Gao, Xingjian, 104
Fenn, John B., 68
Fermi, Enrico, 22, 23, 25, 26
Feynman, Richard P., 250
"First Lady of Physics" Wu, Chien-Shiung as, 247, 251
Fitch, Val L., 51
5-fluorouracil, 209
Fong, Gilbert, 103
force types, 24
Ford, Gerald R., 260
Fowler, R. H., 26
fractional quantum Hall effect (FQHE), 90
Freireich, Emil J., 205, 210, 214
Friedberg, R., 25, 27

Friedman, Milton, 50
Frog (Mo), 152
Furchgott, Robert F., 96

G

Gajdusek, D. Carleton, 50
Gallin, John, 213
Gao, Xingjian, 149, 154
    art exhibitions, 105, 106
    art films, 106
    education, 103
    female–male relationships, 104
    literary career, 103
    Nobel Prize, 103
    plays, 103–106
    writings, 103, 104
Gardner, Frank, 229
Garlic Ballads, The (Mo), 150–152
Ge, Hong, 169
Gell-Mann, Murray, 250
gender discrimination fighter, Wu, Chien-Shiung as, 251
Gengzi Indemnity Scholarship, 118
gestational choriocarcinoma, 202, 209
GFP. *See* green fluorescent protein
glass fibers, 134, 135
Goeppert-Mayer, M., 26
Goldblatt, Howard, 154
Goldin, Dan, 42
Gordon, Robert, 60
Gordon, Sidney Samuel, 196
Gould, Suzanne, 251
gravitational force, 24
"Great Famine", 149
"Great Leap Forward", 149
green fluorescent protein (GFP), Tsien, Roger Yonchien, 120, 121
Guan, Moye. *See* Mo, Yan
Gustaf VI, Adolf, 14
Gustaf, Carl XVI, 34, 49, 65, 81, 95, 108, 125, 142, 159–161

H

Hamaguchi, H., 59
Handbook of Prescriptions for Emergencies, A (Ge Hong), 169
Hankel functions, 134
Hayward, Raymond W., 249
Heisenberg, Werner, 21
hemoglobin, 228
hemoglobinopathies, 231
Herschbach, Dudley R., 58, 60
Hertz, Roy, 205
Ho, Man Wui, Richard, 16
Hockham, George, 134–135
Hong Kong, Tsui, Daniel Chee in, 88, 89
honorary degree
    Chu, Steven, 76
    Li, Choh Hao, 188, 190
honors
    Kan, Yuet Wai, 229–231
    Lee, Yuan Tseh, 61
    Li, Choh Hao, 188, 190
    Mo, Yan, 154
    Ting, Samuel Chao Chung, 46
    Tsui, Daniel Chee, 91
    Yang, Chen Ning, 9
Hoppes, Dale D., 249
Hormone Research Laboratory (HRL), 186–188
hormones, 190, 202, 211
hormone study, Li, Choh Hao, 186, 187
Hua, Loo-Keng, 22
Huang, Kun, 6, 7
Hudson, Ralph P., 249
Hu, Shih, 247
hydrops fetalis, 228–230
hydroxyl group, 169

I

Ignarro, Louis J., 96
induced pluripotent stem (iPS) cells, 231
Institute of Advanced Studies, 23
Institute of Materia Medica, 168
insulin-like growth factor, 187
Intel Science Talent Search, 119
International Telephone and Telegraph (ITT), 133, 136
ion-atom collision, Lee, Yuan Tseh, 60
iPS cells. *See* induced pluripotent stem cells
ITT. *See* International Telephone and Telegraph

J

Jentschke, W., 41
Jiangsu Province, 76
Jinju, 151
Jinling University, 185
Johnson, Frank, 120
Johnson, Lady Bird, 198
Johnson, Lyndon B., 198
J-Particle sign, 45
J/Psi particle, 40–44

K

Kan, Tong-Po, 228
Kan, Yuet Keung, 228
Kan, Yuet Wai
    α-thalassemia, 229, 230
    awards, 230, 231
    clinical training, 228, 229
    discovery, 229, 230
    DNA polymorphism, 228–231
    early life, 228, 229
    education, 228
    family, 228
    honors, 229–231
    iPS cell technology, 231
    molecular genetics of human blood disorders, 231
    polycythemia in hepatocellular carcinoma, 230
    sickle cell disease, 228–230
Kao, Charles Kuen, 8
    Alzheimer's disease with, 137
    career, 133, 136
    early life, 133
    education, 133
    family, 133
    key discovery, 134, 135
    Nobel Prize, 133
    other enabling technologies, 135, 136
Kao, Chun-Hsiang, 133
Karolinska Institute, 264
key discovery, Kao, Charles Kuen, 134, 135
King Faisal International Prize, 9, 10
Koshland, Daniel E., 272
Krugman, Paul, 128

L

La Cantatrice Chauve (Ionesco), 104
Lagosfor, Ricardo, 68
"Land Reform", 149, 151
laser cooling, application, 75
Lasker, Albert Davis, 272, 273, 277
    public service, 274
Lasker Awards, 272–274
    Kan, Yuet Wai, 228
    Li, Choh Hao, 185
    Li, Min Chiu, 202
    Tu, Youyou, 167, 170
Lasker-Bloomberg Public Service Award, 272
Lasker-DeBakey Clinical Medical Research Award, 272
Lasker-Koshland Special Achievement Award in Medical Science, 272
Lasker, Mary, 198, 272, 273, 277
    public service, 274
Laughlin, Robert B., 87, 90, 96
"Law of Parity Conservation and Other Symmetry Laws of Physics, The" (Yang), 6
Lawrence, Ernest O., 248
LeBreton, Pierre, 60
Lederman, Leon, 24, 51
Lee, Chong-Tan, 21
Lee, Mabel, 103
Lee, Tsing-Kong, 21
Lee, Tsung-Dao, 5, 7, 9, 21, 88, 248, 249
    as avid reader, 21
    Yang, Chen Ning relationship with, 24, 25
    director of RIKEN-BNL, 25
    early life, 21
    education, 21, 22
    library, 28, 29
    model, 23
    in National Southwestern Associated University, 22
    Nobel Prize motivation, 21–29
    Physics Scholarship, 28
    postdoctoral mobile stations, 27, 28
    quantum field theory model, 23
    research on astrophysics, 23
    statistical mechanics, 25–27
Lee, Tze-fan, 58

Lee, Yuan Tseh, 58
   awards, 61
   crossed molecular beam apparatus, 60
   dynamics of chemical elementary processes, 58, 61
   early life, 58, 59
   education, 59
   family, 61
   honors, 61
   ion-atom collision, 60
   Nobel Prize motivation, 58–61
Lewis, Gilbert, 186
Li, Ann-si, 188
Li, Arthur Kwok Cheung, 16
library, Lee, Tsung-Dao, 28, 29
Li, Ching-Chen, 73
Li, Choh Hao, 189, 190
   awards, 190
   career, 188
   early life, education and family, 185, 189
   fellowships, 190
   honors and honorary degrees, 188, 190
   Hormone Research Laboratory, 186
   Lasker Award, 185
   momentous achievements, 188, 189
   peptide synthesis, 187
   study of hormones, 186, 187
   University of California, 186
   University of Nanking, 189
Li, Choh Ming, 186
Life and Death are Wearing Me Out (Mo), 151
Li, Jeanette, 203
Li, Jun, 167
Li, Kan-Chi, 185
Li, Ka Shing, 231
Li, Kwok Po David, 97
Limauro, Alvera, 230
Li, Min, 167
Li, Min Chiu, 202
   Albert Lasker Clinical Medical Research Award, 202
   awards, 210
   cancer therapy, 202, 204, 208, 209
   chemotherapy for malignancy, 204–209
   choriocarcinoma treatment, 202–212
   growth of the future scientist, 203, 204
   methotrexate therapy, 203–206, 209, 210
   study on colorectal cancer, 209
Li, Shu-hua, 73
Li, Shu-tian, 73
Li, Tingzhao, 167
Li, Wei-I, 189
Li, Yi-Ying, 118
Li, Yung-Tsuen, 203
Lipscomb, William, 50
Lo, Dennis, 231
logical thinking, concept of, 74
Lo, Mong-hwa, 5
Lowenstein, Louis, 229
Low, F., 26
Lu, Shen Hwai, 185

M
Mahan, Bruce, 59, 60
ma huang, 170
malaria, 168, 171
Ma, Lin, 187
Malmqvist, Göran, 103
Manhattan Project, 248
Mao, Zedong, 103, 168
Marconi's radio transmission, 134
Marshall Scholarship, Tsien, Roger Yonchien, 119
Massachusetts Institute of Technology (MIT), 73, 229
Mayer, J., 26
McDonald, Dong, 60
McFadzean, A. J. S., 229
McNees, Pat, 213
melanocyte-stimulating hormone, 185, 187
metastatic choriocarcinoma, 205
methotrexate, 202–207, 209–211
Mew, Shing Twui, 185
Mills, Robert, 7, 8
MIT. *See* Massachusetts Institute of Technology
molecular genetics of human blood disorders, 231
momentous achievements, Li, Choh Hao, 188, 189
Mo, Yan, 7, 149
   awards and honors, 154
   Big Breasts and Wide Hips, 150, 152, 153

    books, 155, 156
    education, 152
    family, 149
    Frog, 152
    Garlic Ballads, The, 150–152
    legal name as "Guan, Moye", 149
    Life and Death are Wearing Me Out, 151
    Nobel Price, 149–154
    novels, 155
    People's Liberation Army, 149
    Red Sorghum, 153
    Sandalwood Death, 152
    short story and novella collections, 155
    Transparent Carrot, The, 149, 150
Mukden Medical College, 218
Mukherjee, Siddhartha, 211
Murad, Ferid, 96

N

NASA. *See* National Aeronautics and Space Administration
Nathan, David G., 229
National Aeronautics and Space Administration (NASA), 42, 43
National Cancer Institute (NCI), 205, 209
National Center for Biotechnology Information (NCBI), 230
National Central University of Nanjing, 247
National Cheng Kung University, 40
National Southwestern Associated University (NSWAU), 5, 22, 26
NCBI. *See* National Center for Biotechnology Information
Nobel, Alfred, 264–266
Nobel Assembly, 264
Nobel banquet speech, Yang, Chen Ning, 5, 6
Nobel biography
    Chu, Steven, 74
    Tsien, Roger Yonchien, 118, 119, 121
Nobel Committee, 172
Nobel Foundation, 264
Nobel Peace Prize Banquet, 264, 265
Nobel Prize Award Ceremony, 264, 265
Nobel Prize
    award ceremony, 272, 273
    interesting facts, 273

Nobel Prize motivation
    Chu, Steven, 73–76
    Lee, Tsung-Dao, 21–29
    Lee, Yuan Tseh, 58–61
    Tsien, Roger Yonchien, 118–122
non-Abelian gauge fields theory, Yang, Chen Ning, 8
Norwegian Nobel Committee, 264
novella collections, Mo, Yan, 155
novels, Mo, Yan, 155
NSWAU. *See* National Southwestern Associated University
nuclear fission, Wu, Chien-Shiung, 248

O

Obama administration, 76
Obama, Barack, 81
Ōmura, Satoshi, 167
"One Child Policy", 152
Open University of Hong Kong, 154, 231
Oppenheimer, J. Robert, 43
optical communication, 133, 134, 136
optical fibers, 135, 136
optical pumping, 75
Other Shore, The (Gao), 104

P

parity conservation, 24
parity laws, 5, 21
parity of symmetry, 9, 249
Pauli's exclusion principle, 26
Pearson, Olof H., 204
Pei, Chia, 204
Pei, Ieoh Ming, 35, 196
Peking University School of Medicine, 167
People's Liberation Army, 149
peptide hormones, 186
peptide synthesis methods, 187
personal life
    Ting, Samuel Chao Chung, 44
    Yang, Chen Ning, 9
Phase transition theory, 26
Phillips, William D., 73, 75
physics, forces in, 24
Physics Scholarship, Lee, Tsung-Dao, 28
Pines, D., 26

pituitary glands, 185–187, 190
Plasmodium falciparum, 168, 169, 172
Polanyi, John C., 58
polycythemia, in hepatocellular carcinoma, 230
Pople, John A., 96
postdoctoral mobile stations, Lee, Tsung-Dao, 27–28
postgraduate training in physiology, Tsien, Roger Yonchien, 119
Prasher, Douglas, 120–122
Preliminary Explorations into the Art of Modern Fiction (Gao), 103
Prévert, Jacques, 104
Project 523, 168
Prusiner, Stanley, 232
Pui Ching Middle School, 88
pulmonary metastases, 205
Putin, Vladimir, 67

Q
Qing Dynasty, 150
qinghao, 169–171
quantum field theory model, Lee, Tsung-Dao, 23
quantum statistical mechanics, 26, 27
Quantum Yang-Mills theory, 9
quasi-optical waveguides, 135

R
Rabi, Isidor Isaac, 51
radio transmission, 134
Reagan, Ronald, 15
recognitions, Chu, Steven, 76
Red Sorghum (Mo), 153
Relativistic heavy ion collider (RHIC) physics, 25
Ren, H. C., 27
Richter, Burton, 40
RIKEN-BNL, 25
Rink, T. J., 119
Rosenbluth, M., 23
Royal Swedish Academy of Sciences, 264

S
Sandalwood Death (Mo), 152
Saramago, Jose, 96
Sarnoff Robert, 198
Scholarship, Tsien, Roger Yonchien, 119
Schwyzer, Robert, 187
science research, Ting, Samuel Chao Chung, 43
scientific contributions, Wu, Chien-Shiung, 250
Segrè, Emilio, 248
semiconductor lasers, 135
Sen, Amartya, 96
Shangguan, Lushi, 150
Shen, Jun Shan, 27
Shimomura, Osamu, 118, 120, 121, 126
short story, Mo, Yan, 155
sickle cell disease, 228–230
single nucleotide polymorphism (SNP), 230
Sino–Japanese War, 228
SLAC. *See* Stanford Linear Accelerator Center
Sloan-Kettering Institute, 203, 204, 209
Smadel, Joseph, 198
Snow in August (Gao), 105
SNP. *See* single nucleotide polymorphism
Soh, Hsin Pei, 22, 26
solid state physics research, Tsui, Daniel Chee, 90
Soochow University, 21
Soul Mountain (Gao), 104
Space Shuttle Columbia disaster, 42
Standard Model theory, 249
Standard Telecommunications Laboratories (STL), 133
Standard Telephones and Cables (STC), 133
Stanford Linear Accelerator Center (SLAC), 42, 45
Stark, Royal, 89
statistical mechanics, Lee, Tsung-Dao, 25–27
STC. *See* Standard Telephones and Cables
Steinberger, J., 23, 151
Stewart, Thomas Dale, 186
STL. *See* Standard Telecommunications Laboratories
Störmer, Horst L., 87, 90, 96
strong force, 24
subatomic particles, 23, 24
Sung, Joseph, 17, 145

Super Collider Project, 42
Sveriges Riksbank (Sweden's central bank), 264
Swedish Academy, 264

T

Tao, Yuanming, 279
tau (t)-theta (q) puzzle, 23, 24
T. D. Lee Institute, 29
T. D. Lee Library, 28, 29
Teller, Edward, 22
testicular cancer, 204
testicular choriocarcinoma, 209
thalassemia, 228–230
Ting, Kuan Hai, 40
Ting, Samuel Chao Chung, 40
    AMS experiment, 42, 43
    awards, 46
    charm quark, 44
    on collisions of particles, 41, 42
    discovery of J/Psi particle, 40–44
    distinguished awards, 46
    early life, 40, 41
    education, 40, 41
    honors, 46
    J-Particle sign, 45
    Nobel prize, 40, 43–45
    other accomplishments, 44, 45
    personal life, 44
    science research, 43
    universe origin and evolution, 42
Todd, A., 34
Todd, David, 229
Towards a Modern Zen Theatre: Gao Xingjian and Chinese Theatre Experimentalism (Zhao), 103
traditional Peking Opera, 105
Transparent Carrot, The (Mo), 149, 150
Trench, David Clive Crosbie, 35, 195, 196
Tsai, Pei, 58
Tseng, Chao-Lun, 22
Tsien, Hsue-Chu, 118
Tsien, Hsue-shen, 118
Tsien, Roger Yonchien, 118–122
    awards, 122
    early life, 118
    education, 118–119
    green fluorescent protein, 120, 121
    interest in chemistry, 118
    Marshall Scholarship, 119
    Nobel biography, 118, 119, 121
    Nobel Prize motivation, 118–122
    postgraduate training, 119
    Wolf Prize, 118, 122
Tsinghua University, 5, 6, 26
Tsui, Changsheng, 87
Tsui, Daniel Chee, 87
    Augustana College, 89
    awards, 90
    early life, 87, 88
    fractional quantum Hall effect, 90
    in Hong Kong, 88, 89
    honors, 91
    Nobel Prize, 87, 90
    position at Bell Labs, 90
    Princeton University, 90
    in Pui Ching Middle School, 88, 89
    solid state physics research, 90
    University of Chicago, 89
Tsui, Lap-Chee, iv, 97
Tu, Liangui, 167
Tu, Youyou
    Academy of Chinese Medicine, 167
    artemisinin, 169, 170
    awards, 172, 173
    discovery of qinghaosu, 170
    early life, education and enviornment, 167, 168
    experimental human subject, 169
    family, 167
    Lasker Award, 167, 170
    Nobel Prize, 167
    Peking University School of Medicine, 167
    qinghao and eureka moment, 169
    Xiashi Middle School, 167

U

UCSF. *See* University of California at San Francisco
universe origin and evolution, 42
University of California, 186
University of California at Berkeley (UCB), 74

University of California at San Francisco (UCSF), 229, 230
University of Chicago, 89
University of Hong Kong, 22, 228, 231
University of Nanking, 189

V
Varmus, Harold, 230

W
Wandering Spirit and Metaphysical Thoughts (Gao), 106
Wang, Ganchang, 22, 26
Wang, Jeanne Tsun-Ying, 40
Wang, Jwu Shi, 26
Wang, Liming, 172
Wang, Shuangxian, 87
weak force, 24
Weber, G., 41
WHO. *See* World Health Organization
Wilson, E.B., 122
Wolf Prize, Tsien, Roger Yonchien, 118, 122
Wong, C. H., 59
World Health Organization (WHO), 168, 169
Wu, Chien-Shiung, 5, 24
    awards, 250, 252, 253
    beta decay discovery, 249, 250
    cobalt-60, 249, 250
    early life, 247
    education, 247, 248
    family, 247
    "First Lady of Physics", 247, 251
    as gender discrimination fighter, 251
    Manhattan Project, 248
    nick names, 251
    Nobel Prize, 248, 250
    nuclear fission, 248
    scientific contributions, 250
    work in nuclear fission, 248
Wu, Ta-You, 22, 27
Wu, Weishan, 145
Wu, Zhong Yi, 247
Wu, Zuoren, 28

X
Xiao, Shufang, 28
Xiamen University, 5
Xiashi Middle School, 167
Xin, Zhi Ping, 59, 61

Y
Yamanaka, Shinya, 231
Yang, Chen Ning, 23, 88, 248, 249
    awards, 9
    "China Dream", 7
    in Chinese University of Hong Kong, 7, 8
    Doctor of Science degree, 8
    early life, 5
    education, 5
    family, 5
    honors, 9
    "Law of Parity Conservation and Other Symmetry Laws of Physics, The", 6
    Nobel banquet speech, 5, 6
    Nobel Prize, 5
    non-Abelian gauge fields theory, 8
    personal life, 9
    as scientist and artist, 6, 7
    testimonials to, 7
    visit to China, 6
Yang, C. N., 23, 88, 248, 249
Yang-Mills Theory, 7
Yang, Wu-Chih, 5
Yao, Zhongqian, 167
Yeh, Chi-Sun, 22
Yingchun, 151
Young, Rosie, 229
Yuan, Luke Chia-Liu, 248

Z
Zhao, Henry Y. H., 103
Zhejiang University, 22, 26, 247
Zhou, Guangzhao, 27
Zhou, Pei-Yuan, 6
Zhou, Qifeng, 98

www.ingramcontent.com/pod-product-compliance
Lightning Source LLC
Chambersburg PA
CBHW080612230426
43664CB00019B/2865